Energy Revolutions

'The fossil-fuel era was marked by inequality and dominated by oligarchs. David Toke provides a vision of how, in the renewable era, shared ownership of energy could lead to a future of equality and economic as well as energy security.'

—Molly Scott Cato, Professor Emerita of Green Economics, Roehampton University

'Toke doesn't mince words. His title alone says it all. What's needed is not a transition from fossil fuels, as COP 28 meekly requested, but a revolution in how we procure, distribute and use energy. But Toke boldly goes further, calling for a revolution in who gets to develop and own our renewable future. In doing so, he takes on the axis of evil, Reagan-Thatcherism, and their destructive cult of neoliberalism that has shackled renewable energy for decades to the benefit of fossil fuels and their minions.'

—Paul Gipe, author of *Wind Energy for the Rest of Us*

'The future of Humankind depends on the choices we make about future energy systems. Those choices will be ideologically driven: will Western governments go on prioritising obscene corporate profits at the expense of genuine energy security, lower prices and radical decarbonisation? David Toke authoritatively demonstrates how insane that would be – and what brilliant alternatives are already available to us.'

—Jonathon Porritt, author of *Hope in Hell: A Decade to Confront the Climate Emergency*

Energy Revolutions

Profiteering versus Democracy

David Toke

PLUTO PRESS

First published 2024 by Pluto Press
New Wing, Somerset House, Strand, London WC2R 1LA
and Pluto Press, Inc.
1930 Village Center Circle, 3-834, Las Vegas, NV 89134

www.plutobooks.com

British Library Cataloguing in Publication Data
A catalogue record for this book is available from the British Library

ISBN 978 0 7453 4925 1   Paperback
ISBN 978 0 7453 4927 5   PDF
ISBN 978 0 7453 4926 8   EPUB

This book is printed on paper suitable for recycling and made from fully
managed and sustained forest sources. Logging, pulping and manufacturing
processes are expected to conform to the environmental standards of the
country of origin.

Typeset by Stanford DTP Services, Northampton, England

Simultaneously printed in the United Kingdom and United States of America

Contents

Figures

Abbreviations

100%RE	100 per cent renewable energy
BP	British Petroleum
CAUK	Climate Assembly UK
CCGT	combined cycle gas turbine
CCS	carbon capture and storage
CEGB	Central Electricity Generating Board
CfD	contract for difference
DONG	Danish Oil and Natural Gas
DSM	demand-side management
EISRW	Energy Institute Statistical Review of World Energy
EC	European Commission
EPR	European Pressurised Reactor
ETS	Emissions Trading Scheme
EU	European Union
EV	electric vehicle
FPL	Florida Power and Light
GDP	gross domestic product
IEA	International Energy Agency
IRA	Inflation Reduction Act
ITC	Investment Tax Credit
LNG	liquefied natural gas
NGOs	non-governmental organisation
OECD	Organisation for Economic Co-operation and Development
OFGEM	Office of Gas and Electricity Markets
PPA	power purchase agreement
PTC	Production Tax Credit
PURPA	Public Utility Regulatory Policies
PV	photovoltaics
REC	regional electricity company
RPS	renewable portfolio standard
RO	Renewables Obligation

SMR   small modular reactor
SUV   sports utility vehicle

Unless otherwise stated, statistics are taken from the Energy Institute Statistical Review of World Energy (EISRW) 2023.

Acknowledgements

I would like to acknowledge help in various forms from the following people: my wife Yvonne for her help in transitioning my spreadsheets into graphs and for doing the majority of the Index; Philip Palmer for correcting the text; Professor Andrew Stirling for a suggestion that led to this book; some comments on parts of the book from Andrew Tait, Pete Roche, Yvonne Toke and Ed Conduit; and help from Jack Gibbons of the Ontario Alliance for Clean Energy and Bryan Jacob of the Southern Alliance for Clean Energy. Working with Professor Christian Breyer and his team gave me considerable insights and I also thank Michael S. Taylor of the International Renewable Energy Agency (IRENA) for his thoughts. I have benefitted from discussions with my fellow Directors of 100percentrenewableuk Ltd, namely Ian Fairlie, Charmian Larke, Nick Carroll, Keith Kondakor and Herbert Eppel. More generally I would like to thank the MSc students on the Energy Politics and International Energy Security courses who I taught at the University of Aberdeen for the conversations about energy politics.

Preface

This book has been written in the shadow of the massive increases in energy bills for energy consumers. As part of this the fuel poor have been savaged even further. In a bizarre contrast, the already rich oil companies and their shareholders have seen their profits and wealth rocket-boosted by the energy crisis. The post-Covid global energy crisis, the Russia–Ukraine war and the drive to reduce greenhouse gas emissions have all brought attention to the energy industry.

The crisis also coincides with a green energy revolution involving renewable energy and energy efficiency. This book discusses how this revolution can counter the problems of energy insecurity and fossil fuel pollution, especially the climate crisis. It discusses how this revolution is being hampered by the profiteering dominance of the energy corporations. I also outline how people can organise in order to change things. The focus is on key countries in the developed world, in North America, the UK and the EU (with a specific focus on Germany, Denmark and France). This book talks about movements of change expressed in policies and campaigns, especially those that aim to change the dominance of profiteering energy corporations in determining the direction of energy technologies.

This book is set against the backdrop of the 2015 Paris Agreement, signed by almost every country. This commits signatories to keeping emissions down so that post-industrial global temperature increases are below 2°C, and preferably below 1.5°C. A global target of achieving zero greenhouse gas emissions by 2050 is crucial to achieving this objective. In recent years, activists led by people such as Greta Thunberg have blazed a campaigning trail to pinpoint the importance of climate change targets and objectives.

Although many industrialised countries have now set this target for their own countries, this will not be enough to achieve the global target. Developing countries are much less able to achieve the 'zero by 2050' objective. Commenting on the growth of renewables, IRENA says that 'overall deployment remains centred on a few countries and regions, with China, the European Union and the United States accounting for 75% of capacity additions'.[1]

The sad position is that in developed countries the policies needed to achieve even the net-zero by 2050 target are nowhere near being in place. Without the strategies described in this book being implemented there is little chance of these targets being reached.

Chapter 1 focuses on the nature of the green energy revolution and some history of energy and profiteering by the energy corporations that brought us to this point. I then cover decarbonisation energy and show how the threat of climate change can be minimised and energy security increased. This means a focus on moving towards a system based on 100 per cent renewable energy.

In succeeding chapters, we see how far the green energy revolution has gone. Chapter 2 covers the UK, in Chapter 3 there is a discussion about the USA and Canada, and EU countries are covered in Chapter 4. In Chapter 5, I show how people can fight back in the struggle for greater government intervention and energy democracy to achieve the energy transition. I draw conclusions in Chapter 6.

There are some key themes. First, there is the way in which laws and government policies have been fashioned to serve the interests of the multinational oil and gas corporations and the big electricity and gas utilities. Given their own reliance on fossil fuels, their interests have lain in delaying or distracting from the task of meeting climate change targets such as those set by the Paris Agreement. They have seen their interests as maintaining their fossil fuel, and sometimes nuclear, status quo assets.

Second is how energy revolutions have begun to gather force. I talk about the green revolutionaries that have started the process and the role that green energy technologies are playing in energy transformation. I discuss how this can lead to a 100 per cent renewable energy system.

Third is the means of transforming energy demand through demand-side efficiency and through the electrification of transport and heating.

Fourth, 'leaving things to the market' has been shown by the energy crisis to be a failure. The market did little to look after national interests, especially in Europe, which has been most hard hit by increases in natural gas prices.

Fifth, I address the role of state intervention to ensure cost-effective deployment of renewable energy and energy efficiency. I talk about community ownership in promoting innovation in green energy developments, the role of public ownership in energy, the role of major state-owned companies and municipal owned energy companies.

Sixth, I look at the importance of energy democracy, that is how decisions about technology are made by people rather than merely corporations. Currently the energy companies control the information necessary to make government policy decisions, and invariably these decisions are dominated by the energy companies themselves.

1

Pathways to Energy Revolutions

A GROWING REVOLUTION

In March 2009 I attended a UK government meeting of UK businesspeople associated with the development of ports for the offshore wind industry, listening to the UK energy minister, Mike O'Brien, talking about the government's plans for offshore wind. He exclaimed, 'We need to bring about a revolution in the way energy is produced. ... Imagine you are pin-striped revolutionaries in the spirit of Che Guevara on the Sierra Madre.'[1] His appeal to the revolutionary spirit was noble. However, the real green revolutions came many years earlier in the form of grassroots campaigners for wind and solar power in Denmark and Germany in the 1970s to 1990s.

Since 2009, there have certainly been the beginnings of a major industrial shift towards renewable energy across the world. In the UK the share of wind and solar photovoltaics (PV) as a proportion of UK electricity has increased from 3 per cent in 2009 to 24 per cent in 2021.[2] In the USA it increased from 2 per cent to 13 per cent in 2021.[3] That, of course, is just electricity.

It is important to understand the challenge that lies ahead. After all, 82 per cent of world energy was recorded as coming from fossil fuels in 2022.[4] On the other hand, the rate of change towards renewable energy is accelerating. In the world, from 2009 to 2021, total energy from non-hydro renewable energy (mainly wind and solar) increased threefold.[5] This total now easily surpasses nuclear electricity production. Green energy technologies are now seen as central to future economic development, as witnessed by the movements in Europe and the USA for a 'Green New Deal'. This movement has not yet achieved all its aims, but the direction is clear.

We need to see these developments as a sharp corrective to those shameful Western politicians who say we should not take radical action because it all depends on what China does. The point is that the developed world needs to take the lead given its great wealth and technological enterprise. It can offer finance and expertise to poorer developing countries.

For the purposes of this book there is the assumption that 'renewable' refers mostly to wind power, solar power and water-sourced renewable energy, including hydro, tidal and wave power. Geothermal energy may also be seen as renewable. Some biomass, such as biogas from wastes, may also be renewable. However, there are big question marks over how low carbon and sustainable other biomass resources may be. Certainly, we cannot rely on future growth of these if we are to deal with the climate crisis. The reason that growth in renewable energy deployment is becoming exponential is because the cost of renewable energy has been declining so much.

Denmark is the world's leader in wind and solar, as a proportion of total electricity generated. The Danes derived almost 60 per cent of their annual electricity from wind and solar energy in 2022.[6] From an energy revolution point of view, Denmark is important, because it was in that country that rural craftspeople and anti-nuclear energy campaigners began to develop modern wind turbines from the mid-1970s. They were aided by green energy activists from California, who backed wind power in the 1980s and bought wind turbines made in Denmark. As markets for wind power grew, it was a trend that led eventually to the growth of what is now the giant multinational wind power industry. The Danish wind power company Vestas is still the world's leading wind turbine manufacturer. Those local Danish people who first championed community- and farmer-owned wind power were true revolutionaries![7]

Besides Denmark, Germany was full of revolutionary green energy campaigners. In 1990 they succeeded in persuading the government to organise a system of funding wind power, and these 'feed-in tariffs' enabled the renewable energy producers to receive a guaranteed payment for their electricity sold to the system. In 2000 this was extended to solar PV on the back of green energy campaigning. I played a small role in all of this, helping the UK play catch-up with the revolutionaries. In 2007, I wrote a report which helped to spearhead the campaign to press the UK government to institute feed-in tariffs for solar PV.[8] This was successful and, alongside similar efforts in several other European countries, a substantial market opened up for solar PV.

The early support for wind and solar power that appeared across the West was attacked as being expensive. However, the markets that developed as a result meant that the wind and solar industries grew quickly. The costs of renewable energy plummeted, and today renewable energy is much cheaper than either fossil fuels or nuclear power. If things had been left as the anti-renewable incentive campaigners wanted, then of course the renewables industry would never have taken off. The world would be in a parlous position in terms of surviving the fossil fuel price spirals that we see in cycles (in both oil and natural gas price crises). Our ability to deal with the climate crisis would be almost destroyed.

Many people in Africa still rely on unsustainable locally collected biomass use for their essential energy needs.[9] Hence we still need to shift from unsustainable biomass consumption to reliance on renewable energy sources. Besides increasing deforestation, biodiversity loss and greenhouse gas emissions, this bad biomass use is also damaging people's health through smoke inhalation.

Africa tends to have more solar resources than the more economically rich North yet has so far lacked the resources to develop solar power. It used to be said that Africa needed to be helped to survive on conventional fossil fuels, since they were affordable compared to expensive new renewable energy. However, things have changed, and now wind and solar power should be cheaper than fossil fuels. They certainly should be in the South, but shortages of capital to buy the equipment and a lack of expertise in new renewable energy sources is slowing the transition to sustainable fuels in Africa.

Here the role of the state should be obvious, in expanding the foreign aid that is given by developed countries to help start renewable energy industries in Africa. A key aspect of this is developing human capital, and Northern universities should be funded to train Africans in renewable energy expertise in order to plan, manufacture and service renewable energy in Africa. African governments should be encouraged to sponsor university students in Northern universities to study renewable energy. This leads me to justify the Northern-centric nature of this book by saying that up until now energy innovation has been mainly done by the North (for good and often ill), and so the focus here is on considering what changes the North can bring that will help itself and the rest of the world become more sustainable.

PROFITEERING

The incredible scale of profits by the oil companies reflects the absence of democratic intervention in the energy system to protect poor consumers, who have been struggling with high energy prices since 2021. In 2022 the major oil and gas companies made particularly strong profits. Exxon made $55 billion dollars, Shell $40 billion, Chevron $36.5 billion and Equinor $55 billion, among other eye-watering numbers for several other oil and gas majors.[10] In fact, Equinor is the only company that will return a lot of its profits to the state, in its case the Norwegian state, since Norway owns around two-thirds of the shares in Equinor. Even this number of course has been reduced by the continued free market blandishments to reduce the state's share of the ownership of Equinor. The result of this free market ideological pressure is that the wealthy, who own the shares, get richer at the expense of ordinary people.

Analysis of World Bank data suggests that the profit levels obtained by oil companies are not just an aberration of the recent energy crisis. According to research conducted by Professor Aviel Verbruggen from the University of Antwerp, and stated by the UK's *Guardian* newspaper: 'The oil and gas industry has delivered $2.8bn (£2.3bn) a day in pure profit for the last 50 years'. Verbruggen is quoted as saying: 'It's real, pure profit. They captured 1% of all the wealth in the world without doing anything for it.'[11]

Some states have imposed 'windfall' taxes on profits on oil and gas revenue from extraction on their land, but in practice this is only a small part of the profits of these companies. Such taxes tend to be limited in actual effect by the 'investment' allowances that the companies can have to avoid paying tax. I am not suggesting here that we can simply nationalise oil assets to solve this problem – the compensation required to do so would be massive. However, there are three lessons. First, this private profiteering emphasises the folly of those countries, including the UK and Norway, who have sold off oil and gas assets that were owned by the state for ephemeral short-term gains that have been massively overshadowed by profits that have been earned since. Second, the notion of a country being 'secure' because of fossil fuels produced on its territory is illusory if the fuels are sold on a world market where domestic consumers have to pay the same crisis-bloated prices as everyone else. Third, as the renewable energy revolution gathers pace, we need state intervention to ensure that the benefits of lower-cost green energy supplies go to the consumer and not the energy corporations.

This necessity for greater state intervention to guard against corporate profiteering from renewable energy was underlined when, in 2021–3, natural gas prices rose in Europe. Electricity from gas-fired power plants became very expensive. All prices on the wholesale power markets in Europe are set at the level of the highest-priced generation, which in this case was from gas-fired power stations.

However, in theory a lot of renewable energy was cheaper because they were developed based on contracts issued by governments. These contracts guaranteed minimum payments to the renewable operators for each unit of energy generated. However, the cost savings from cheaper renewables were very often not passed onto consumers. Often big corporations who owned renewable operators or had long-term contracts with them were able to sell on the power they produced at the (very high) wholesale power prices.

As discussed in Chapters 2, 4 and 5, this is at least partly an outcome of neoliberal ideology applied to power markets in the UK and the European Union (EU). We need to extend government intervention and elements of state ownership of the retail energy supply sector to ensure that the consumer, not the big corporations, benefits from cheap renewable energy. We need to stop the profiteering that occurs with fossil fuels being extended in the future to renewable energy.

In the USA, corporate monopolies, who control much of the USA's electricity sector, have regulated profit levels. However, this does not guard against profiteering since this makes their corporate investments profitable before they even start generating. This encourages them to maximise investment in their own power plant and keep out independent generators of electricity markets. In the USA, shutting off energy services to consumers reached record levels in 2022. However, at the same time, monopoly electricity utilities, which generally have high profit rates, were seen to be lavishing rewards on executives and investors at the same time as punishing poor consumers. Over four million cut-offs were carried out in the first ten months of 2022.[12] The UK also saw a controversy in early 2023 as it appeared that increasing numbers of people were being forcibly put onto expensive meters despite being on lists of vulnerable people.

ExxonMobil has produced its own self-serving projection of global energy in 2050 in which oil and gas use expands beyond present levels. This can be seen as part of a campaign to deflect efforts to stop further oil and gas drilling. ExxonMobil claims that emissions can be reduced in such a scenario through use of carbon capture and storage and hydrogen produced from natural gas.[13] Yet these technologies are expensive, polluting and inefficient compared to use of renewable energy and energy efficiency.

ENERGY CRISES

Energy crises have revolved around the dominant fuel of the time. As more and more of the world has become dependent on fuels with a global supply market, there has been a growing trend towards global energy crises. In the Western world in the eighteenth to nineteenth centuries, coal arose as the dominant fuel (displacing wood), originally to power increased demands for consumption of clothes and other items. After the mid-nineteenth century there was a rising suburban middle class who wanted greater mobility to escape from the industrialised cities.[14] The first motor vehicles were powered by coal-fired steam engines.[15] However, the marriage of oil and the internal combustion engine proved to be the winning formula thereafter to satisfy the new middle classes. Gradually oil penetrated more and more energy sectors – not just transport, but also industry, heating and even, by the 1960s, much electricity production. However, these voracious demands produced such rapid increases in demand that eventually supply problems emerged. This translated into the oil crises of the 1970s.

There were oil price spikes in 1973, 1979, 2008 and 2011. Although political factors are often said to be the cause of energy crises, the context is that of an increasing imbalance of supply compared to rising demand. Nuclear power was touted as an alternative to oil. Some, often worried by connections between civil and military uses of nuclear power and the risks posed by radioactivity including nuclear waste, wanted an alternative that used the renewable processes of the Earth itself: the wind, the sun and the water. These 'alternative energy' activists (in the 1970s and 1980s) were seen as fringe oddballs by the energy mainstream. Today their vital role in developing niche renewable energy technologies and markets is airbrushed out of history since it contradicts the idea that big capitalism solves the big problems.

As Karl Marx pointed out, the social forces that power revolutions first form as seeds in the previous revolution. In this case a nascent pro-renewable energy movement formed in the relative undergrowth of a fossil fuel-dominated world. The world responded to the oil crises by replacing oil with either coal or gas, and generally paying a bit more attention to energy efficiency as a response to higher energy prices. Yet the green energy revolution began to put down some roots. Hence, when further oil crises occurred in 2008–2011, the incentives for renewables powered wind and solar technologies forwards to bigger markets. Tremendous cost optimisation occurred. Now the latest energy crisis has supercharged the renewables revolution.

Most recently there has been a natural gas crisis in many parts of the world. This crisis is multifaceted and is certainly made worse by Russia's invasion of Ukraine. However, it is important to note that the global natural gas market is very different to that of oil because, compared to oil, a much smaller proportion of natural gas is shipped (as liquefied natural gas (LNG) as opposed to being piped). Hence, natural gas prices are quite different in different regions of the world, whereas there is much less difference between oil prices in different parts of the world. This global LNG market has been under increasing pressure in recent years as demand from the East, especially China, has increased. The consequence is that natural gas-importing countries such as European states are paying higher prices for natural gas. North America was less affected by this crisis because it has self-sufficiency in natural gas supplies. However, the USA is now exporting so much natural gas to Europe that it may end up subjecting its own citizens to higher global natural gas prices.

A supreme irony of the political battles over energy technologies is that those opposed to the promotion of renewable energy have championed fossil fuels as providing energy security, compared to 'intermittent' renewable energy. Yet recent history should suggest the opposite, in that renewable energy, when given support from governments based on fixed price contracts (as in the UK), are actually a hedge against high fossil fuel prices. Moreover, the renewable energy supplies are domestically based and therefore much more subject to domestic control compared to the globalised oil and natural gas markets and the dominant profiteering oil and gas companies. Of course, energy efficiency is the most secure of all energy options.

A BOTTOM-UP REVOLUTION FOR THE ENERGY SYSTEM

Decentralised energy and energy efficiency was conceptualised in technological terms by Amory Lovins, who wrote a seminal paper for the journal *Foreign Affairs* in 1976. He contrasted what he saw as the 'hard' energy path – a centralised, high-energy, high-waste/pollution, nuclear power-dominated future that appeared to be the consensus – with what he saw as a 'soft' energy path. This soft path involved decentralised energy technologies and a great emphasis on reducing energy demand through energy efficiency. It was a very new approach at the time.[16]

The revolutionary soft path also includes decentralised technologies for using the energy. Energy is supplied to provide energy services, so obviously the thing to do is to reduce the amount that is needed to provide the services. Greens are keen on using heat pumps to provide energy to buildings, since pumps are not just decentralised but are also highly efficient producers of energy. Compared to fossil fuel technologies (e.g. natural gas boilers) heat pumps reduce energy consumption by around 70 per cent. Heat pumps do this by using inputted electricity to suck heat out of the environment, whether air, ground or water. The energy can be used for heating or cooling. Heat pumps are a hybrid technology involving both energy efficiency and direct renewable energy. I recall a leading academic expert (Professor Nick Eyre of Oxford University) on the subject of energy efficiency calling for the British 'Insulate Britain' protestors to think of campaigning for heat pumps rather than insulation.[17] On the other hand, analysts such as Jan Rosenow and Sam Hamels say that the energy efficiency of buildings (mainly through insulation) needs to be improved dramatically, otherwise the costs of supplying the heat through heat pumps 'may exceed the limits of what is cost efficiently achievable'.[18]

Another possible candidate for energy efficient heating is infra-red heaters. These would not supply hot water, however. In addition, the jury is still out on how much energy savings they produce and whether they could provide as good a service as a conventional radiator system.

Electrification is an essential part of progress towards an energy efficient economy and will dramatically reduce energy demand. Electric vehicles (EVs), alongside heat pumps (and using electric stoves for cooking), provide by far the biggest part of this general move towards electrification. Electric cars more than double the efficiency with which energy is used compared to traditional petroleum power motor vehicles. Carbon Brief reported that '[i]n the UK in 2019, the lifetime emissions per kilometre of driving a Nissan Leaf EV were about three times lower than for the average conventional car, even before accounting for the falling carbon intensity of electricity generation during the car's lifetime'.[19]

As the proportion of energy, especially electricity derived from renewable energy, increases, so the carbon-reducing bonus of replacing internal combustion engine vehicles with EVs becomes bigger and bigger. Hence electrification is very important because, besides leading to considerable improvements in energy efficiency, it also allows a vector for most renewable energy technologies which generate electricity. That will make it a lot easier to supply what is left through renewable energy.

EVs allied with residential solar PV and batteries have the potential, alongside utility-scaled renewable energy and battery developments, to become substitutes for a significant part of the power station capacity that exists today. EVs, heat pumps, other electricity using equipment and batteries can be digitally connected. This can reconcile fluctuating renewable energy sources with fluctuations in demand for energy. For example, people can store electricity produced from renewables either from their rooftop solar panels or through the grid in their home battery – or even the battery in their electric car. This will be when power prices are cheap. Later in the day they can draw power from their batteries to provide their energy needs rather than take energy at a time when it is more expensive.

We need to adopt the revolutionary way of discussing energy systems as ones that include the way energy is used and the technologies that use energy, not by the conventional division between consumption on the one hand, and generation, transmission, distribution and supply on the other hand. Oil and gas energy systems are, by tradition, described as consisting of 'upstream' activities involving energy extraction and 'downstream' activities such as distribution and supply. Sometimes 'midstream' activities are identified, which include fuel processing such as oil refining. But of course, this structure applies less to renewable energy, where the 'extraction' is done directly by the generating machines. Renewable energy and nuclear power are mainly generated as electricity, and electricity can be divided into generation, transmission and (more local) distribution, and supply. Formulations of energy systems which see consumers as passive consumers of energy supply are dated.

The revolutionary approach to energy means that, for example, buildings need to be designed with an energy system mentality, as part of the energy system. This includes technologies that reduce energy use to provide a given service and help the system to balance fluctuating renewable energy supplies. This also includes designing buildings to use as little energy as possible in the first place through using thicker walls and roofs and more insulation materials. In addition, buildings should have designs that encourage easier cooling and heat recovery. There needs to be a major retrofit programme of insulation for existing buildings and new buildings need to be designed to require no energy input with solar PV and batteries fitted.

Greens will emphasise urban planning that gives priority to bicycles and walking rather than motor vehicles. Electric cars are part of the solution, but at the same time, cars in general are also a large part of the problem. Reducing local pollution levels, and returning space to people and nature and away from motor vehicles, is an important aspect of decentralisation.

Greens will also emphasise incentive structures that give opportunities to local communities rather than merely corporations. Hence, green movements have promoted incentives such as feed-in tariffs for renewable energy that are open to anyone. Feed-in tariffs offer fixed payments for each unit of energy production that are guaranteed over a set number of years. This is as opposed to incentive systems, such as the US-style production (now called investment) tax credit that favours large companies and which are difficult for small or community schemes to access.

In sum, the revolutionary approach that comprises green energy features three key aspects. First, a focus on energy efficiency and renewable energy as the path to not just decarbonisation but as green ends in themselves. Second, it is important to emphasise the decentralising of energy institutions and the empowering of people. This includes transforming homes into green energy centres, and involves local cooperatives, community ownership of renewable energy schemes and progressive municipal ownership of energy companies. Third, at a macro state level, it involves changing the regulatory and incentive structure to favour renewable energy, energy efficiency and green approaches to transport.

IMPLEMENTING A HEAT PUMP PROGRAMME

The 'no-brainer' first step should be to immediately set building regulations that ban use of fossil fuels in new buildings. Heat pumps appear as the chief best option for heating and air conditioning, while conventional electric heating and infra-red heaters are also possible options. Objections to a fossil fuel ban on new buildings based on preserving 'freedom of choice' are misguided, since people do not have freedom of choice about heating technologies when they move into new buildings. We would not expect 'freedom of choice' about whether buildings should obey safety rules! Allowing the possibility of 'hydrogen heating' through 'hydrogen ready boilers' is just an excuse to keep people hooked on using natural gas.

The more difficult part of the energy transition is retrofitting existing buildings with a non-fossil fuel alternative. This is perhaps best done by giving mandates to municipal authorities and electricity utilities. Municipal authorities should have the powers and duty to plan and install heat networks serviced by large-scale heat pumps (including the ability the raise money for this).

Electricity utilities need to be given obligations to install a given number of heat pumps every year (or possibly infra-red heaters if appropriate) in individual buildings not connected to heat networks. Doing it this way may be the cheapest option compared to funding each individual through grants, although these could also be made available. It is best if the costs of doing this are spread among all consumers by way of a levy on bills rather than expecting it all to be financed by individual consumers. All of this will be expensive but spread over a quarter of a century it should be easily affordable.

An important element of a heat pump programme will be to ensure that electricity prices are kept down by ensuring that renewable energy generators are paid high, fixed, long-term prices for their generation. Public ownership of the retail energy supply is likely to help produce this outcome (see discussion in Chapters 2 and 5).

ELECTRIC BATTERY PLANES?

Currently the commercially available lithium batteries are too heavy to allow their use in all but small aeroplanes going short distances. However, recent laboratory-based advances in battery technology mean that electric air travel may make rapid strides in the coming couple of decades possibly by 2040. For example, the Argonne Laboratories have reported a breakthrough using so-called lithium air batteries. This could lead to an increase of up to four times the amount of energy held in a battery of a given weight.[20]

This will also make EVs a lot better, and hydrogen vehicles are unlikely to make as big an impact as once thought (perhaps not at all). There will still be some niche uses for hydrogen, but they are likely to be restricted to industrial energy use, where electricity is not desirable, and also for storing energy. The same breakthroughs that help pave the way for better electric battery aircraft also pave the way for cheaper, more efficient, longer-range and less material-intensive EVs. This is discussed in Box 1.2.

ELECTRIC VEHICLES

The cost of manufacturing EVs is falling rapidly and the range of new EV models is increasing. A report in the *Guardian* said: 'The falling cost of producing batteries for electric vehicles, combined with dedicated production lines in carmarkers' plants, will make them cheaper to buy, on average, within the next six years than conventional cars, even before any government subsidies, BloombergNEF found.'[21] The fears of pressures on lithium and copper resources for EVs are likely to be misplaced. In the case of lithium, battery recycling (which has been given a boost by Biden's Inflation Reduction Act (IRA)) combined with increasing efficiency of lithium use in new designs for batteries will ease pressures on lithium resources. In the future, EV batteries are likely to need much less copper. According to Elon Musk, 'Tesla expects that by moving to a 48-volt system for the secondary battery – the smaller battery used to power functions like lighting and wipers – in future EVs, it will be able to cut the need for copper to one quarter of current levels.'[22] Similarly, the need for other in-demand heavy metals such as cobalt will be reduced by a mixture of alternative technologies and greater efficiency of material use.

The number of people switching to EVs around the world is accelerating. According to International Energy Agency (IEA) figures, global car sales rose nearly fivefold between 2019 and 2022.[23] People have stopped talking about when peak oil production will occur, and instead the talk is about when peak demand will set in.

Of course, it is the case that even though in just a few years most new cars will be EVs, this will still leave a very large residue of old petrol or diesel vehicles. Yet these may disappear more quickly than might otherwise be thought. There are two reasons for this. First, campaigns for cleaner air in urban areas are likely to lead to clean air zones and also to fossil fuel vehicle scrappage schemes, which will reduce the attractiveness of fossil fuel cars.

Eventually, in any case, we shall see a tipping point where the number of petrol and diesel refuelling stations falls because of their lack of profitability. Of course, none of this will distract from the success of movements to campaign for more investment in mass transit systems such as trams. Movements to reduce car traffic need to succeed in creating car free or reduced car access city areas where cycling and walking are given priority over motor vehicles.

HOW THE ESTABLISHMENT HAS BEEN PROVED WRONG ABOUT RENEWABLE ENERGY

The rapid development of renewable energy is taking off both inside and outside the West – in China, India, Brazil and so on – which is a trend that defies the earlier predictions of some great establishment energy sages. Writing in 2005, Vaclav Smil, for instance, opined that 'in the world's affluent countries the dependence on fossil fuels has risen to such a high level – averaging now in excess of 4 kW/capita – that there is no alternative technique of non-fossil energy technology that could take over a large share of the supply we now derive from coal and from hydrocarbons in just a few decades'.[24]

Smil's prediction seems to be heading for trouble. The declines in the cost of wind and solar over recent years has been dramatic. According to the independent Lazard Business Consulting Group, solar and wind power had by 2021 become the cheapest supply options for new electricity generation of any type.[25] As can be seen in Figure 1.1, the share of renewables is increasing much more rapidly than what might have been predicted even 20 years ago.

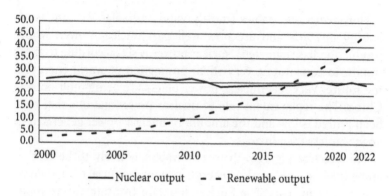

Figure 1.1 World nuclear versus non-hydro renewable output this century (exajoules)

Source: Energy Institute Statistical Review of World Energy (EISRW) 2023.

However, now it seems that just taking over 'a large share' of the energy supply will not be enough. As it is, the goal set by the Paris Agreement of 2015 of rapidly reducing greenhouse gas emissions, basically to zero by 2050, seems a long way off. Carbon emissions were stable in 2018–22 after centuries of increases in consumption of fossil fuels. Period.

Renewables and energy efficiency are likely to do a lot more in the future than merely keep emission levels stable. That is, provided greenwashing delaying tactics by the fossil fuel industries and their corporate allies are defeated. Non-hydro renewables have overtaken nuclear power in terms of share of world energy consumption. Nuclear power's share of world electricity consumption in 2021 declined by over 40 per cent from a peak in 1996.[26]

The rate of renewable energy expansion is escalating. In December 2022, Faith Birol, the chief executive of the IEA, said '[t]he world is set to add as much renewable power in the next 5 years as it did in the previous 20 years'.[27] In 2022 fossil fuels still dominated the world's energy supply, with oil gas and coal constituting around 82 per cent of total primary (that is raw) energy consumption (including both electricity and non-electricity energy). However, the growth of renewables is phenomenal, as can be seen in Figure 1.1.

Non-hydro renewables (mostly wind and solar PV) are accelerating upwards and are now well past nuclear power, whose production has stagnated. Nuclear power is set, according to these trends, to constitute a very small proportion of world energy in the future. Very few nuclear power plants are being built outside China. The replacement rate of nuclear power stations that have been retired is low and the amount of nuclear power generated in 2022 was less than in 2000.

The accelerating growth in renewables is likely to be further embedded in the world's energy system because of a continued decline in costs. The Rocky Mountain Institute, for example, predicts that this fall in prices will continue. As an article in *Cleantechnica*, which summarises Rocky Mountain Institute thinking, puts it:

> Renewable energy costs have fallen, and are projected to keep falling, because these technologies are riding 'learning curves.' For every cumulative doubling of the deployed tech, its cost declines by a quantifiable percentage that varies by technology. Over the past 40 years, the average learning rate has been 20% for solar and 13 per cent for wind.[28]

Indeed, if recent growth trends in renewable energy continue, then sustainable renewable energy sources (mostly wind and solar PV) will make up 100 per cent of world energy consumption (all energy, not just electricity) by the year 2050. This is shown in Figure 1.2. Nuclear power is not shown here because its contribution is likely to be trivial. Based on trends over the last ten years, nuclear power would be projected to supply only around 3 per cent of world energy in 2050.

There is a consistent trend in the last ten years of world growth in renewable energy (mostly wind and solar power) of 12.6 per cent per year (according to the 'Renewables Consumption' section of the EISRW database used for Figure 1.1). By contrast, the total primary energy consumption (that is, all energy, not just electricity) is showing an average growth of 1.4 per cent per year over the previous ten years (see the 'Primary Energy Consumption' section of the EISRW database). If we assume that these growth

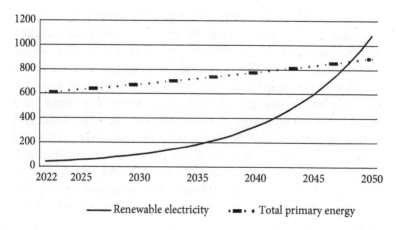

Figure 1.2   Future world renewables and total world energy consumption using last ten years of trends (exajoules)

Source of trends data: EISRW 2023. Data on electricity generation from biomass drawn from Ember.

rates continue into the future, we can see (in Figure 1.2) that by 2048 renewable energy (including 2022 levels of hydroelectricity consumption) surpasses 100 per cent of total world energy consumption. Note this is based on *all* world energy, not just electricity. This is shown in Figure 1.2 (line with dots is total energy and the continuous line is renewable electricity). In calculating this projection the total amount of electricity generated from biomass has been excluded from this figure as there are doubts whether much of this can be counted as sustainable.

Of course, it may well be that the recent growth rates of renewable energy may slow, but any such movement is likely to be counterbalanced by a slowdown and, well before 2050, the beginning of a consistent contraction in world primary energy growth. There are good reasons to argue that total world primary energy demand will stop increasing as much as assumed in Figure 1.2. This is because the likely accelerating use of EVs and heat pumps in the future will dramatically reduce energy consumption in the transport, heating and cooling sectors.

China's energy growth rates have slowed greatly in recent years and the Chinese economy is looking more like a 'mature' industrial economy where there is less spending on providing basic infra-structure. Energy consumption is likely to be in significant decline well before 2050. Many Western countries are already seeing declines in energy consumption and this trend will accelerate with the adoption of electrification technologies (mainly heat and EVs, including electric battery aircraft). These trends may more than offset increases in energy consumption coming from India and Africa. In those countries, increases in energy consumption will be much more moderate than during previous waves of industri-alisation. They will be adopting the new electrified technologies and also powering their development with renewable energy and building industries that are much less energy intensive than has been the case previously in the West or East.

My assumptions will be controversial with many analysts. However, the assumptions that are frequently used by analysts – for example, that the use of fossil fuel-powered aircraft and motor vehicles will continue indefinitely – seem to me to be much less likely than electric battery technologies becoming universal in these sectors. Heat pumps may now seem expensive to retrofit in old buildings, but in decades hence the processes of doing this will become refined and industrialised. Indeed, the likely exponen-tial future growth in these technologies will expand the market for renewables greatly since most sustainable renewable energy is produced in the form of electricity. Looking at the picture in this way, it is plausible that the world can be served completely using solar and wind power (and maybe also other renewables) by 2050. Experts schooled in the analysis of the past and present are often left blindsided by technological revolutions into predict-ing slow change. I am sure the establishment predictions are going to be wrong in the case of energy. The green energy revolution is coming and it will happen with such a speed that it will take people by surprise. There is a glimmering of understanding emerging in the energy establishment about these trends. Residues of IEA bias against renewable energy still shine through as represented by its treatment of data. The IEA's contribution is discussed in Box 1.1.

Political opposition, especially opposition from the fossil fuel industry, could slow the advance of renewables. Grid constraints to connecting renewables need to be cleared. In addition, a lot

Box 1.1 The IEA: boosting nuclear energy and fossil fuels at the expense of renewables?

The IEA was formed in 1975 as part of the rich Western nations' response to the fact that developing countries had seized control of oil production from the Western oil corporations and wanted a decent price for their product. Many people assume that the IEA is an independent body which undertakes consensus-driven analysis. However, the IEA is not independent. It is run by a cabal of mainly Western governments. The Secretariat is based in nuclear-dominated France.

The IEA's widely used statistical database gives around three times the arithmetic energy value to a kWh of nuclear power generation compared to a kWh of solar PV or wind power generation. It achieves this statistical effect by counting all the energy produced by a nuclear power plant, even though around two-thirds of the energy produced is thrown away in waste heat (into rivers and seas).[29] Its statistical method downplays the colossal wastage of energy involved in producing nuclear power. It also downplays the immense wastage of energy used to generate electricity from fossil fuels. So, the useful energy contributions of both nuclear power and fossil fuels are exaggerated by this method relative to wind and solar power. Of course, no waste heat is associated with wind farms or solar farms, so this is not counted.

As such, the fact that renewable energy has overtaken nuclear power in the amount of electricity generated in the world today is obscured in the IEA data. Indeed, ExxonMobil uses the same statistical method – one which conveniently minimises the numerical contribution to world energy supply of wind and solar compared to oil and gas. This harms public debate.

The IEA has, in the past, had a reputation of miserably failing to predict the growth of renewable energy year after year. That is despite charging large amounts of money for its self-proclaimed 'authoritative' forecasts in its 'World Economic Outlook' (WEO). *PV Magazine* commented in 2020 that:

> [E]very year since 2002 the WEO solar forecasts have been wrong, and, in almost every year, from the very first year of the forecast. This is because solar has grown far faster even in the first year of the forecast period than most of the forecast's project for 25 years later. That's not just wrong. That's catastrophically and laughably wrong. Year in and year out.[30]

However, the good news is that recently the IEA's projections for solar and renewables have improved. They are now extolling the virtues of how renewable energy is rapidly expanding. Less happily, it appears to be slow to change the ludicrous way it counts nuclear generation compared to wind and solar power.

depends on how much governments are prepared to support consumers to change their heating and air conditioning systems to use heat pumps. All of this can only happen with consistent campaigning from civil society groups to push for practices and policies that support the energy transition. Although big corporations will inevitably play a leading role in this transformation, we must ensure that we do not allow them to switch from profiteering with fossil fuels to profiteering from renewables.

PUBLIC OWNERSHIP AND COMPETITION

Many aspects of the energy sector are natural monopolies and therefore the benefits of private competition are not possible in such sectors. The transmission system and local distribution systems are natural monopolies. Local distribution can be (and currently often is) owned by municipal authorities.

I favour a pragmatic socialist approach that promotes public ownership where it can achieve better outcomes than private ownership – and this occurs much more often than assumed by dominant Western ideas. Public ownership can be increased in ways which will boost competition. One example of this is the establishment of state companies to develop renewable energy alongside existing private companies. Existing publicly owned companies can be given more ability to further the objectives of deploying renewable energy and energy efficiency. Often they are stopped from doing this by rules that favour the private sector, but such practices are clearly anti-competitive, especially when, as is often the case, they are denied the ability to take advantage of financial incentives that are available to private corporations (as has been the case in the USA). Public ownership also has an important role in delivering services in parts of energy systems where competition is itself either impossible or inefficient. It may be especially relevant to the retail electricity supply sector. There is more discussion of this in Chapters 2 (on the UK) and 5.

However, in the energy sector there have sometimes been efforts to invent competition, for instance in the retail energy supply sector, where there is little possibility for useful competition. Here the costs (to the alleged 'competing' energy suppliers) of trying to market to many small consumers are more than any gains that can be made by the companies. Moreover, many consumers are not sufficiently interested in playing the market, which further reduces the utility of the so-called competitive arrangements. Indeed, why should consumers have to spend their time chasing deals on the internet? The majority would prefer not to do this, and it leaves the suppliers using loopholes in regulations to penalise loyal consumers. This is demonstrated in the UK by the fact that the majority of consumers do not bother to switch suppliers.[31] Bringing in retail energy supply into public ownership should be cheap for the state to achieve since the companies involved have few tangible assets.

A crucial shortfall in the liberalised market arrangements is the difficulty in organising effective demand shifting through time-of-use tariffs that coincide with the periods during which wholesale electricity is traded (i.e., varying electricity prices according to half-hour trading). This is important to balancing fluctuating renewables. It is no good if regulators simply assume that competing electricity suppliers will offer half-hour time-of-use tariffs as an option for the consumer. This has to be organised through offering incentives to all consumers to engage in demand flexibility; something that is unlikely to happen if only some suppliers offer an effective demand-side flexibility service.

In addition, savings from cheap renewable energy supplies need to be passed through to consumers. Liberalised markets in Europe have allowed energy generators to make excess profits, although renewable energy is generated using relatively cheap power purchase agreements. Public ownership of the retail energy supply can remedy these problems.

Talking positively about public ownership has been generally unfashionable since the 1970s and the inception of the Thatcher–Reagan era in the West. Neoliberalism is often used on the left to generally refer to the post-Reagan–Thatcher politics of reducing state involvement in the economy and reducing taxes for the better off. However, this general definition merges two subtle but meaningful differences between European and American interpretations of so-called neoliberalism. Essentially, the European approach is more concerned with avoiding monopolies, while US conservatives seem to be much more content with allowing corporate monopolies or near monopolies to get on with things without state regulation.

In Europe, following on from 'ordoliberal' approaches in post-World War II Germany, a notion of neoliberalism has been followed that emphasises an 'economic constitution that in its entirety is tuned to upholding competition in the face of uncompetitive interests'.[32] This has relevance to the energy sector as discussed at much greater length in Chapters 2 and 4, covering the UK and the EU respectively. The Europeans have gone in for liberalisation of energy markets more thoroughly than is the case with the USA.

This European neoliberalism, which is what I mean when I refer to neoliberalism in this book, has meant that artificially competitive markets have often been created in places which, though they may have been privately owned before, did not involve competition. This includes the retail energy supply sector, where we have seen the invention of what turns out to be illusory competition. Of course, the promotion of real competition is unobjectionable, especially in the electricity generation sector. Helping independent renewable energy generators compete with corporately owned power stations has been essential. Indeed, as mentioned, adding state-owned renewable energy development companies into the mix acts to further increase competition in what has become an international market.

What I call the neoliberal approach is also popular among some Democrats in the USA. Conservatives in the USA, however, are generally content to let the corporations run things with less obedience to the notion of achieving 'fair' competition.[33] I discuss the modalities of neoliberalism as it relates to energy more in Chapter 2 on the UK and Chapter 4 on Europe. The problem of neoliberal policies pricing cheap renewable energy according to expensive power from gas is also discussed.

HOW DOES HYDROGEN FIT INTO
THE GREEN ENERGY REVOLUTION?

Some have argued that hydrogen-powered vehicles are the way ahead, assuming that the hydrogen will be produced using renewable energy. However, hydrogen-powered cars are likely to use energy a lot less efficiently than electric cars, because of losses in the electrolysis of water to produce the hydrogen and also in the fuel cells which are powered by the hydrogen. It has been thought that hydrogen would be better suited for trucks and other large vehicles, but even here advances in electric battery technologies are making this less likely. Even worse, it is feared that far from using green hydrogen the hydrogen will be derived from fossil fuels, with or without some carbon capture and storage.

Claims by the fossil fuel industry that they can use existing natural gas networks to power homes using hydrogen are rightly seen, by greens, as greenwash. The claims are a self-serving attempt to prolong the use of natural gas. Low-carbon hydrogen production and delivery systems are yet to be seriously developed, and anyway they will use energy a lot less efficiently than heat pumps. Figure 1.3 compares the amount of energy input needed to generate the same useful heating output from (1) a heat pump powered by renewable electricity and (2) a boiler fuelled by green hydrogen. Heat pumps generate much more useful heat than the electricity they consume because the electricity is used to draw heat from the environment for heating purposes (like a reverse fridge). Use of green hydrogen does not do this but it loses a lot of its energy during conversion from renewable electricity in the hydrogen. This implies that we would need a great deal more renewable energy to power a 'hydrogen' economy compared to one which uses direct electricity as much as is practically possible.

So-called blue hydrogen produced from fossil fuels involves extracting hydrogen from natural gas and capturing carbon dioxide in the process. Yet it is not clear whether this can be done cost-effectively without wasting a lot of energy that is consumed in making the hydrogen and letting too much carbon dioxide escape. Of course, hydrogen can, and should, be produced from renewable energy through the electrolysis of water, which leaves behind only oxygen as a by-product.

'Green' hydrogen needs to be used only for essential purposes, for example for storing renewable energy or for some industrial purposes for which electricity is not desirable. It should not be squandered in the provision of heating or cooling services. Heat pumps will only need around a quarter of the renewable energy

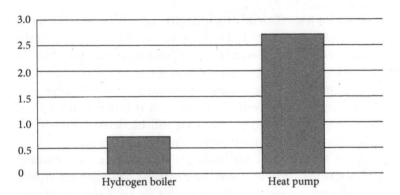

Figure 1.3   Heating output using 1 kWh renewable energy from hydrogen boiler and heat pump

Note: Vertical axis represents kWh of heating output. Assumptions: heat pump coefficient of performance 2.7, efficiency of electrolysis process for electricity to hydrogen is 80 per cent and boiler efficiency of 90 per cent.

to provide the same level of heating or cooling compared to using green hydrogen. Rates of heat pump installation vary in different countries, as can be seen in Figure 1.4. The coldest countries, such as Norway and Finland, appear to be associated with the highest level of heat pump installations. The UK has just about the lowest installation of any Western economy. This position is entrenched

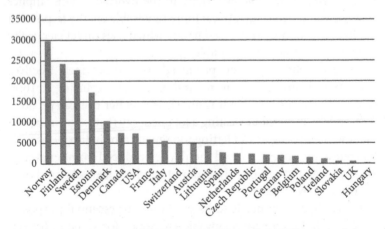

Figure 1.4   Heat pump installations per 100,000 people in different countries

Sources: Adapted from data compiled by Josh Jackman and published by the 'The Eco Experts',[34] European Heat Pump Association, Canadian Climate Institute, Ecotechnica, the US Air Conditioning, Heating and Refrigeration Institute, Worldometer.

by all sorts of nonsense about heat pumps spread by some people with interests in maintaining the natural gas industry.

Hydrogen can be used to power motor vehicles, but again (as in heating) this is a much less efficient way of using renewable energy, this time compared to EVs using batteries. An EV is likely to use renewable energy more than twice as efficiently as a hydrogen-powered vehicle. This is because the fuel cells used in hydrogen vehicles will be only around 50 per cent efficient, and on top of that energy will be lost turning the electricity into hydrogen through electrolysis. By contrast, EVs turn most of the electricity input into usable energy. Consequently, it is sales of EVs that have taken off around the world and not hydrogen vehicles. The growth in EV sales can be seen in Figure 1.5.

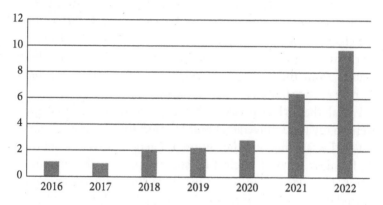

Figure 1.5   World electric car sales, 2016–22 (millions)

Source: IEA, 'Global EV Outlook 2023', www.iea.org/reports/global-ev-outlook-2023/trends-in-electric-light-duty-vehicles.

Another possibility for hydrogen use is in air travel. Green hydrogen can be made using electricity from renewable energy to electrolyse water, although there are issues about how the hydrogen can be stored for sufficient capacity for long-haul flights. A second option is to use net zero carbon synthetic fuels. This also involves the use of hydrogen as well as carbon dioxide sucked in from the atmosphere. This process would use up a lot of renewable energy. By 2050 it may be that most air travel is done by battery-powered electric planes.

A HISTORY OF THE PRESENT

If we are to understand how energy revolutions occur and how technology develops and changes, we need to think about energy history. Most energy histories assume that it is about technological progress. But technologies change in relation to social forces as well as developing technologies. Dealing with the climate crisis certainly turbocharges the march to build renewables. However, there is more to it than this.

To understand the significance of the green energy approach we need to perform a 'history of the present' – a discussion that avoids conventional understandings which smooth out all of the historical wrinkles to show how the present preoccupations have been seamlessly developed over time. This concept was coined by the French political sociologist Michel Foucault.[35] At the beginning of the chapter, I mentioned the activist revolutionaries in Denmark, Germany and California who created markets for renewable energy from the 1970s, so the green revolution started before climate change became a well-known issue.

I recall that as late as 1987, when I was doing a master's degree, I went to my local university library to find some references for the course I was doing on 'The Future of Energy'. I asked the librarian on the desk where I might find some papers on the 'greenhouse effect'. He then picked up an index dealing with gardening and looked up greenhouses. Seriously, it happened! It was not because the librarian was ignorant, at least for his time, it was just that the issue which initially was called the greenhouse effect rather than climate change, or most recently the climate crisis, did not even begin to hit the public imagination until at least the following year. Important policy commitments did not start to emerge before the UN Rio Conference on Environment and Development in 1991.

The modern environmental movement, concerned not only with protecting some areas of nature but also with combatting industrial pollution, emerged after World War II. This was especially in the wake of controversies about the effects of open-air atomic bomb tests and then the impact of pesticides, about which Rachel Carson wrote.[36] Up until then, as David Nye put it: 'There was an unmistakeable emphasis on progress, on subduing nature, and on using nature to achieve both.'[37]

The modern green energy revolution itself earns its genesis from the energy politics which followed on from the 1973 oil crisis. This crisis showed the nascent green movement that humans were depleting and polluting natural resources at an unsustainable rate, producing alarming levels of different types of pollution. There had to be a shift towards conservation of resources and the use of renewables rather than radioactive nuclear power.

By 'green groups' or the 'green movement', I refer to environmentalist campaigning organisations such as Greenpeace, the Sierra Club, Friends of the Earth and a range of nature protection organisations and green parties, the last of which coalesced in the decade following the 1973 oil crisis. After World War II, nuclear power had developed, often with links with the nuclear weapons programme, and greens were peace campaigners opposed to nuclear weapons. Also, green ideology promoted support for decentralised technologies. These included locally oriented solutions based upon the theory that local people could live sustainably if they could see that they were not exceeding the carrying capacity of the land upon which they lived.

Many anti-nuclear activists became active in promoting renewable energy. It is notable that the two European countries with the strongest early (in the 1980s and 1990s) anti-nuclear movements, Denmark and Germany, also had the strongest pressures to develop wind and solar power. California also had a strong anti-nuclear movement, which led to an important, early state-sponsored renewable energy programme.

The rise of climate change as an issue led the green movement away from anti-nuclear campaigning as its first priority and towards tackling climate change. However, it has left a residue of suspicion and continued opposition to nuclear power among most green groups. The fact that nuclear power has proved, and is proving, so expensive to develop has given many green non-governmental organisations (NGOs) reason enough to argue that wind and solar power be given preference in investment programmes. A series of accidents at nuclear power stations – Three Mile Island in the USA in 1979, Chernobyl in the Soviet Union in 1986 and Fukushima in Japan 2011 – has further tarnished the image of nuclear power.

Some, such as Professor Andrew Stirling (Sussex University), argue that there is still a clear link between military and civil uses of nuclear power. The same companies that make nuclear weapons or nuclear submarines are lobbying governments to give them more money to promote civil nuclear power. Professor Benjamin Sovacool (also University of Sussex) argues that countries which prioritise renewable energy rather than nuclear power achieve greater levels of carbon reduction. Certainly, it does appear that the strength of renewable energy programmes in different countries depends on their reliance on and interest in civil nuclear power. This appears to be the case in the countries shown in Figure 1.6.

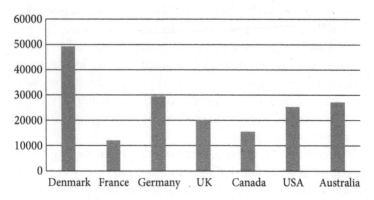

Figure 1.6 Consumption of non-hydro renewables per person in selected countries, 2022, in TJ per person per year

Sources: EISRW Energy 2023 and World Bank population figures for 2021.

Figure 1.6 uses countries featured in this book for a comparison of development of non-hydro renewables (i.e. ones that have been added in recent decades). Countries which have abandoned or do not use civil nuclear power, namely Denmark, Australia and Germany, have the most output of (non-hydro) renewables per head, while the countries that do support nuclear power – the UK, the USA and France – have the least renewables outputs per person. This is especially the case in France with its very large, but now seriously declining, nuclear power industry. Given the difficulty and cost in delivering new nuclear power, it does seem foolish to pin hopes on an unlikely nuclear revival when resources can profitably be ploughed into renewables.

REVOLUTION VERSUS CORPORATE POWER

Mainstream governmental energy policies have signalled support for some important parts of the green agenda. Several Western governments, including the USA, the UK and the EU, have agreed to set a 'net zero' target for greenhouse gas emissions by 2050. There are programmes in place to support the continued expansion of renewables and some support for electrification strategies involving EVs and heat pumps.

However, this apparent consensus hides a struggle between the decentralised green revolutionaries and corporate interests. The corporate interests want to salvage as much as they can of their dominance and they often drag conservative politicians along with them. The policy preferences of the energy corporations often rely heavily on support for fossil fuel-based solutions such as capturing carbon dioxide emissions from power plants, carbon capture and storage (CCS), or turning natural gas into hydrogen (blue hydrogen) and capturing the carbon as a by-product. In fact, CCS projects have a poor track record, especially in the power sector, where a meta study found that there was not a single successful project.[38] At best, CCS is likely to capture no more than 90 per cent of carbon emissions, which is itself inadequate, but in reality the experience so far has fallen well short of this.

On the other hand, schemes for 'blue' hydrogen have been criticised as being very poor in terms of cutting greenhouse gas emissions.[39] Inefficiencies in converting natural gas into hydrogen mean that the emissions released when the hydrogen is generated will offset much of the gain in capturing the carbon dioxide during the steam-reforming process that is used. The efficiencies of the system may be increased, but in doing so, costs will increase. This on top of the fact that the source is natural gas, with potentially volatile prices.

Nuclear power is also a feature of the corporate approach to decarbonisation, and this coincides with the preferences of the energy corporations and their emphasis on centralised power stations and the centralised despatch of energy as the main solution. Yet despite being offered considerable subsidies, new nuclear power is hardly being delivered at all in Western countries. Hopes of a 'nuclear renaissance' have been expressed since energy prices started rising in 2004, but this has never materialised. The few projects that are being built are hopelessly over budget and hopelessly delayed.

Among the problems facing efforts to develop new nuclear power plants, there are four big issues. First is the fact that nuclear power plant designers have incorporated safety features designed to minimise the consequences of nuclear accidents, but in doing so the plants have become much more complicated and difficult to build without great expense. A second reason is that large construction projects of whatever type, at least in the West, tend to greatly overrun their budgets.[40] In the West, improvements in health and safety regulations to protect construction workers have no doubt played a part in this. A third factor is that, in the West at least, the cheap industrialised labour force that dominated the industrial economies of the past and which could be used to develop nuclear programmes (in the way that France did in the 1980s) has ceased to exist. A fourth factor is simply that renewable energy technologies, especially wind and solar power, can be largely manufactured offsite in a modular fashion and their costs have rapidly fallen, leaving nuclear power increasingly uncompetitive.

Talk of small modular reactors (SMRs) is now common, but such concepts have been tried in the past and do not provide a credible way forwards for civil nuclear power. Small reactors already exist in the form of small, pressurised water reactors (PWRs) used in nuclear submarines. However, at such a small size, PWR technology is extremely expensive – which is the reason PWRs have been built at a much larger scale. There are 'advanced' modular reactor designs, but these consist of reheated notions of concepts that were abandoned 60 years ago. Nuclear power plants have to be large to absorb the costs of necessary safety measures and achieve economies of scale for the reactors and their steam turbine sets. There is no reason to believe that small reactors will be successful in the future. This is outside of a fantasy world where hundreds of cities demand their own modular reactors with their citizens willing to pay virtually any price for them. Steve Thomas has commented on this phenomenon.[41]

Renewable energy can be seen as a solution that will reduce consumer costs and be implemented much more quickly than nuclear power, something that became clearer in Europe during the recent natural gas price crisis. However, it also became clear that savings were only passed through to consumers when there were the right contracts and right type of public regulation of the energy supply industry. I discuss how this can best be done in Chapter 2 on the UK and Chapter 5.

100 PER CENT RENEWABLES

In recent years, there has been increasing interest in the possibility of achieving net zero carbon emissions involving all energy, not just electricity, being sourced from 100 per cent renewable energy. Support for this seems to come from places where there is an aversion to investment in new nuclear power plants. This includes several states in the USA and in Australia (where new nuclear power plants are actually proscribed by law). The Australian Market Operator has produced a 'roadmap' for the transition to obtaining 100 per cent of the electricity supply from renewable energy.[42] See a discussion of how South Australia's supplies of solar and wind power are supplying 100 per cent of electricity in Box 1.2.

The German government has committed to obtaining 100 per cent of its electricity from renewable energy by 2035.[43] Germany has had a policy of phasing out nuclear power since soon after the Fukushima accident in 2011. Mark Jacobson, a Stanford University professor, has popularised the notion of all energy (not just electricity) being supplied by 100 per cent renewable energy using the abbreviation WWS – Wind Water and Sunlight.[44]

There are certainly enough renewable energy resources available to provide the world with 100 per cent of its energy from renewables. According to Jacobson, solar PV theoretically exists, over land alone, to provide the world's 2050 end use power 640 times over.[45] The sea-based resources (yes, there are already floating solar farms!) are around three times larger than the quantity of land-

> ### Box 1.2 100 per cent renewables in South Australia
>
> Australia has immense renewable energy resources, especially solar PV and wind power, and in 2022 it produced the fourth highest amount of solar PV generated per person in the world (after Denmark, Sweden and Norway).[46] South Australia is among the biggest states in the country for solar and wind production, to such an extent that Giles Parkinson reported in *RenewEconomy* that in December 2022, 'South Australia has just chalked up what is undoubtedly a world first – a run of more than 10 consecutive days over which the average production of wind and solar accounted for 100 per cent of local demand'.[47]
>
> There are lots of regions where renewable energy has supplied 100 per cent of electricity, but they have involved a lot of hydro electricity. However, in this case (South Australia) there was no hydroelectricity in play, just fluctuating wind and solar production. There is an increasing number of batteries being built to soak up excess generation. In the future there will be longer and longer periods of 100 per cent renewable energy production. South Australia has a separate grid to the rest of Australia, although it can take imports and export power.
>
> Parkinson further comments:
>
>> During the 10-day period that averaged 100 per cent wind and solar, South Australia only had a minimum amount of gas generation – not for energy needs, but for 'system strength' and other vital grid services. ... This need is likely to be redundant in a few years when the new link to NSW [New South Wales] is built, and as more 'grid forming' inverters that mimic the properties of coal, gas and hydro, are constructed. ... Then, we will see an even more remarkable landmark – one that so many people still insist is not possible – a gigawatt scale grid running for a period of time on wind and solar only, with no fossil fuel generation at all.[48]

based resources.[49] The amount of wind resources that are available depends on what minimum wind speeds are assumed, and this is a moving target since wind turbines are being made more efficient and cheaper, meaning that lower wind speeds become more economical. Wind resources are somewhat lower than those available from solar PV but, according to Jacobson, this would still be nearly 30 times higher than total world energy consumption in 2050.[50] In the case of the UK, for example, Wind Europe has estimated that the gross potential from offshore wind is around 60 times the current total electricity consumption of the UK.[51]

I justify the aim of achieving 100 per cent of all energy from renewables using three reasons. First is the argument that renewable energy sources, by definition, do not run out – that is at least as defined earlier as comprising energy from wind, sun, water sources, geothermal and perhaps some biomass wastes. A second justification is that these sources are (usually) also non-polluting as far as the atmosphere is concerned. Related to this is a third reason, which is simply that, given the priority given to eliminating carbon dioxide emissions from energy use, there is no alternative. That is because nuclear power and fossil fuel CCS do not represent realistic alternatives.

Wind and solar power are advancing quickly in terms of increasing capacity and power production (although not yet quickly enough), while energy production from nuclear power is stagnant.[52] On the other hand, as mentioned earlier, there are practically no CCS power stations in operation in the world, and none so far that can be regarded as being successful. It may be plausibly argued that better schemes will emerge in the coming years, but the fact still remains that most investment is in renewable energy. Little investment is going into nuclear power or CCS. There are very good economic reasons for this given the relative cost of both nuclear power and fossil fuel projects involving CCS, compared to wind power and solar power, which, as discussed near the beginning of this chapter, are much cheaper.

We often hear of 'breakthroughs' in the search for the ultimate elixir of nuclear fusion. However, if the evidence is inspected clearly, as opposed through rose-tinted hallucinations, we are light years away from nuclear fusion being a reality. Claims of creating net energy in nuclear fusion tests are only correct if we ignore the vast amount of energy inputted to create the reactions in the first place. Conventional nuclear power is in technological decline.

As shown in Figure 1.1, conventional nuclear production has been stagnant this century, and around the world the replacement of old nuclear reactors is occurring at a very slow rate. Once the current spurt of labour-intensive industrialism peters out in China, their drive in building nuclear power will fade, leaving nuclear in decline. Nuclear power's usefulness in a renewables-dominated system is small. That is because, for financial reasons let alone technical ones, nuclear power cannot afford to switch on an off as required to help balance renewable energy.

Hence there arises a simple and obvious argument: there is no alternative to planning for 100%RE systems. Countries such as the USA and the UK, which are committed to achieving net zero greenhouse gas emissions by the year 2050, may be taking a big risk if they base their systems planning on two types of energy sources – nuclear power and fossil fuels with CCS – which are not going, in practice, to provide much of a contribution towards energy supplies in the future. It may be much cleverer to base planning on the basis of a 100%RE outcome. Failure to plan for a 100%RE system, or something very close to it, will produce one of two outcomes. The first is that we may miss our carbon emission reduction targets by a substantial margin. The second is that failure to set an objective of 100%RE and, importantly, to plan for it effectively, may produce a system that is not properly balanced, leading to systemic disruption.

One obvious barrier to 100%RE is the extent to which there can be sufficient renewable energy deployed to meet all the energy requirements of a particular country. However, given the figures discussed earlier for world renewable energy resource availability, in theory at least there should easily be enough capacity available, at economic cost, to meet energy demand. Wind and solar power, again as discussed earlier, are now often, if not usually, the cheapest new options available for energy generation. Political barriers exist, and these will be harder to overcome in some places rather than others. However, if 100%RE is seen as an important political objective, then these barriers are more likely to be overcome.

In addition to ensuring adequate uptake of renewable energy resources and energy efficiency, another leading concern with the practicality of a 100%RE system is balancing the variable nature of wind and solar production with demand profiles for energy. This implies a need to store renewable energy in some form or other.

There are three types of balancing that are necessary in a 100%RE or near 100%RE system. First is intra-day balancing to cope with solar and wind output being different to the level of energy that is demanded at any one time. This involves (1) greater management of demand, through incentives to companies and domestic residents to consume energy at some times when renewable energy is immediately available rather than at other times when it is less so. This is likely to involve use of EV batteries (as well as a growing share of grid-scale batteries) to take surplus power at some times and also put it back into the grid when there is a power shortage. (2) Gas-fired generation to provide electricity when renewable supply is low. The gas used could be green hydrogen (produced by electrolysing water with renewable energy) or green synthetic fuels (again produced from renewable energy) which could be used on conventional gas engines or power plants. Other plausible techniques for managing short-term fluctuations in supply include using pumped hydro projects. Thermal stores (large hot water tanks) could be used to avoid using large-scale heat pumps feeding heat networks when there is not enough renewable energy.

The second type of balancing demand is between seasons. There need to be systems of storage to cope with the variability of wind and solar power between seasons. A third type of balancing is inter-annual balancing, which is concerned with balancing years in which renewable energy output may vary. It is much too expensive to invest in batteries to deal with inter-seasonal or inter-annual balancing because they would be used very rarely. There are two ways of dealing with this longer-term balancing. One is 'overbuilding' renewable energy, that is installing more renewable energy capacity than would be needed to supply enough energy in an average year. This can avoid a lot of the need for energy storage. However, there would still need to be enough energy storage to cope with weeks when there was little wind or solar production. One study concluded that in a 100%RE UK electricity system a 15% 'overbuild' would reduce the need for storage to 30 days' worth of energy storage.[53] An alternative to overbuilding renewable energy is simply to build much larger energy stores, although this would require several months' worth of storage to be available.

Green hydrogen could provide a carbon neutral fuel that could be stored in underground caverns for long periods (in much the same way that natural gas is stored today). Low-NOx hydrogen generators are being designed in order to reduce the amount of nitrogen oxide that is produced by the generators.[54] Biogas from waste product may be sustainable, but the resource is limited by the waste resource itself. Geothermal energy could also help store energy in a renewable form. One interesting innovation in geothermal technology is the Eavor scheme. This promises the possibility of providing renewable energy on demand through a geothermal system that avoids the use of 'fracking' techniques.[55] Compressed air has already been used as a storage medium in electricity systems and this could be deployed to store renewable energy.

Another solution could be to use renewable energy to produce carbon neutral methane or methanol. The production of green synthetic fuels (requiring not just hydrogen from hydrolysis but also carbon sucked from the air using renewable electricity) is rather less efficient compared to producing green hydrogen. However, set against this, the (underground) storage of hydrogen takes up a lot of space for any given weight, and also holds less energy for a given weight compared to other substances such as synthetic methane or synthetic methanol.[56] These fuels have a much higher energy density compared to hydrogen. They are likely to be much cheaper to use compared to green hydrogen. An additional advantage from using such synthetic fuels is that they can be used in existing gas engine- or gas turbine-based power stations.

Building underground gas storage systems and gas power stations has done a lot already. Although much gas generation capacity is needed to balance a 100%RE system (using green hydrogen or green synthetic fuels), gas generation equipment is a minute cost (capacity for capacity) compared to, say, nuclear power. The cost of the equipment may be over 20 times cheaper compared to the costs associated with recent attempts to build nuclear power plants in Western countries. In a UK-oriented study, it was found that developing a 100%RE system is likely to be much cheaper than a pathway of meeting net zero targets by 2050 using nuclear power and fossil fuels with CCS.[57]

One interesting approach is to imagine providing 100 per cent of energy from renewables in the context of a globally interconnected electricity system. This would have the advantage of connecting areas where it is daytime with areas where it is night, as well as more and less windy zones. In recent decades, new engineering solutions for interconnection involving high-voltage direct current have emerged. These allow the possibility of (economically) transmitting electricity across thousands of miles while minimising electricity losses. A group of researchers has modelled the possibilities for a global system to provide 100%RE. They concluded that, compared to systems that are not globally interconnected, a globally interconnected system would reduce storage costs for 100%RE by 50 per cent and reduce the costs by 20 per cent.[58]

CONCLUSION

For so long the public has been told that energy security requires promoting oil and natural gas above renewable energy. Yet globalised energy markets mean that giving power to the oil and gas companies does nothing for energy security and provides great profits for oil and gas. A revolution was started by a few Danish and then German and Californian green activists in favour of green energy, and it has grown. Now even multinational energy corporations have realised they must take notice of this revolution, even though it may mean attempting to divert it, to prop up their own fossil fuel or dinosaur-nuclear interests.

Energy security has always tended to be a slogan that has been shaped in the interests of the big energy corporations. Yet this has been made threadbare during the recent energy crises. The recent crises are not just about the Ukraine war, although that has certainly made things a lot worse. It is about the sheer volatility of the fossil fuel energy system, from which the big energy corporations profit, and for which ordinary consumers pay.

Increasingly, governments in places such as the USA and the UK have agreed to targets to reduce carbon emissions. Green campaigners advance the strategic solutions of (1) energy efficiency, especially through electrification of heating and cooling using heat pumps; (2) renewable energy, principally including wind and solar power; and (3) better grid interconnections and storage and flexibility solutions to help balance the grid.

Yet the solutions promoted by governments all too often help the energy corporations make money out of preserving their own interests rather than supporting the cleanest and most cost-effective solutions. Many billions are being poured into technologies such as CCS in fossil fuel power plants, 'blue' hydrogen produced from fossil fuels and even SMRs. Yet these schemes are much less cost-effective and certainly much less innovative than renewable energy or energy efficiency technologies. There seems to be little thinking about the consequences of rolling out a super-costly system of nuclear power across the world, giving dictatorial, repressive regimes access to nuclear technology which can be used as cover for the development of nuclear weapons, or at best (and this is pretty bad) seeing their energy systems fall under the control of Russian nuclear agencies.

Insufficient government activity is directed towards the most likely solution – that is, a system that will trend towards high reliance on renewable energy and energy efficiency – rather than CCS power plants or nuclear power. Plans need to be developed to form a 100%RE system, but this is simply being ignored in favour of fantasy plans that favour the energy corporations. One thing we do not want to do is to create a new situation where big corporations charge consumers excessive amounts for supplying increasingly cheap renewable energy. The importance of increasing state intervention and of deploying techniques of energy democracy to achieve these ends are discussed in Chapter 5.

2

How Neoliberalism Wrecked UK Energy – and How to Turn Things Around

A PROPHECY?

On 19 October 1992, Michael Heseltine, the secretary of state for trade and industry, made a statement to the House of Commons about the large-scale closures of coal mines that were to take place. This was a consequence of the privatisation and liberalisation of the energy industries. In the House of Commons debate that ensued, Neil Kinnock, who had until three months previously been leader of the Labour Party, said: 'The Government's attitude is now based on the historic error of depending on cheap gas and cheap imported coal, which is an invitation to companies and countries that owe nothing to this country to hold our country to energy ransom in the future.'[1]

At the time, this may have sounded, to many, like a Member of Parliament, representing a traditional coal mining area, clutching at straws to defend an outmoded source of energy in the shape of coal mined in the UK. But in the autumn of 2022, these words looked almost prophetic, in the light of the world natural gas price spike, giant profits being made by multinational oil and gas companies and actions by Russia to squeeze Europe by cutting off supplies of natural gas.

In this chapter, I examine the extent to which Kinnock was being prophetic, in the sense that privatisation and liberalisation of energy was responsible for harming UK energy security. To do this, I examine the history of energy after World War II, focusing especially on how energy was privatised and liberalised. Then it will be possible to draw up a 'counterfactual' picture of what the UK energy position might have been, had energy not been privatised and liberalised. To what extent might energy security have been better without privatisation? And to what extent might greenhouse gas emissions have been differently impacted?

Policy has changed tremendously since the early 1990s. The Climate Change Act (the result of campaigning by Friends of the Earth) was passed in 2008. This led to the establishment of the Committee on Climate Change and a target of reducing green-house gas emissions by 80 per cent by 2050. In 2019 this target was updated to involve a 'net zero' emissions target for greenhouse gas emissions to be achieved by 2050. Friends of the Earth won a High Court ruling against the government in July 2022 for not giving enough information showing how the required reductions in carbon emissions were going to be achieved.[2]

Since 2011, the emphasis on renewable energy has made a sig-nificant difference in lowering carbon emissions. Also, emissions have been significantly reduced because of a fall of 15.3 per cent in UK total energy consumption in that period. However, as discussed in Box 2.1, the large reductions in carbon emissions recorded by the UK since 1990 have much to do with structural factors, espe-cially deindustrialisation.

The high level of reliance on oil and natural gas for transport and heating has left the UK vulnerable to variations in oil and gas prices. Indeed, the UK became even more relatively exposed to volatile global gas prices because reliance on natural gas increased in this period. Prices increased from 2006 and then went to crisis levels from 2021. The changes in natural gas prices (using a constant value of money) can be seen in Figure 2.1.

The changes in the consumption and production of natural gas can be seen in Figure 2.2. Consumption (and production) of natural gas increased in the years following privatisation as more natural gas was produced to supply gas-fired electricity generation. However, this led to the depletion of natural gas reserves, and a decline in production set in after the early years of this century. Increasingly, gas consumption outpaced production of natural gas.

Gas prices increased throughout the European market for natural gas after 2005. The effect of the gas price increases seems

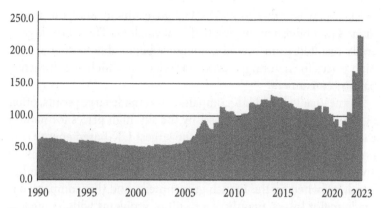

Figure 2.1 Changes in UK gas prices since 1990 expressed as an index where 2010=100

Source: Domestic Energy Prices, UK government energy statistics, www.gov.uk/government/statistical-data-sets/monthly-domestic-energy-price-stastics.

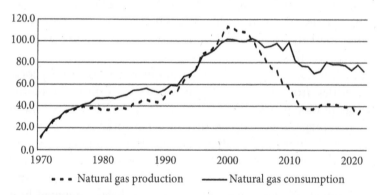

Figure 2.2 UK natural gas consumption and production since 1970 (billion cubic metres)

Source: EISRW 2023.

to have been to moderate demand for gas. Natural gas prices in the UK vary in responses to European gas prices. European gas prices can be seen in Figure 4.1. Oil prices for the UK, on the other hand, will reflect global oil prices. Gas prices remain regionalised (e.g. in Europe) compared to oil prices, which are more globalised. For more discussion of this, see Chapter 4.

As discussed in Chapter 1, the big oil and gas corporations made massive profits out of the 2021–3 energy crisis. British Petroleum

(BP), which was owned by the UK government until the 1980s, made $14.7 billion in the first half of 2022 alone. These profits generally resulted from the oil companies being able to charge very high prices for natural gas as well as oil, costs which are ultimately paid by consumers.[3]

Domestically based UK companies also made large profits, often out of high electricity prices set by the sky-high prices for natural gas generation. Together, five of the largest UK-based energy corporations – Centrica, E.ON, RWE, Scottish Power and SSE – made around £8 billion in the first part of 2022.[4] EDF UK, on the other hand, is owned by the French government, and this company has made major losses, mainly as a result of problems with its nuclear power fleet. Oil and gas executives may claim that continuing to drill for oil and gas protects UK interests, yet the experience of the energy crisis undermines such claims. Regardless of the existence of oil and gas generation from UK, British energy consumers had to pay prices set on world markets. The fact that oil and gas for British markets were sourced from British waters made little or no difference to this fact.

Much of early electricity and gas industry was owned by local authorities from the end of the nineteenth century until the nationalisation of gas and electricity in 1947–8.[5] They were central in developing the electricity distribution network.[6] Nationalisation was a lot to do with centralisation, rather than bringing firms into public ownership. Centralised and ever larger coal-fired power stations were favoured rather than smaller, locally based units, even though the local generators may have been combined heat and power operations, with good total energy efficiency of electricity and heat production combined. This ideology of centralisation also favoured a nuclear power construction programme. This was launched in the 1950s, initially as a side effect of nuclear weapons production, and in the 1960s an increasing number of oil-fired power plants were built.

So-called town gas, produced from coal, was replaced with natural gas sourced from the North Sea from the late 1960s. This involved changing domestic gas appliances, but it represented a major improvement in the reduction of various types of pollution.

PRIVATISATION AND LIBERALISATION
EQUALS NEOLIBERALISM

Following the energy crises of the 1970s (involving coal strikes and international oil price spikes), and with the political economy drifting to the right, support for privatisation increased. Initially, when Margaret Thatcher became prime minister in 1979, this was concerned with reversing the previous Labour government's nationalisations. Then the agenda moved on, with energy featuring prominently. The sale of shares in BP (a company originally formed to exploit oil in Persia before World War I) began in 1979. In that year the UK government ceased to be a majority shareholder and the remaining government-owned shares were sold off in 1987. Similarly, the British National Oil Corporation, which had been created in 1975 to exploit North Sea oil for the British state, was privatised in 1982.

British Gas, a company formed following the nationalisation of gas in 1947, was then privatised in 1987. At the time this was framed as an effort to spread popular ownership of shares through the 'Tell Sid' (about the share offer) campaign, although in reality these shares where quickly bought up by corporate interests. There was criticism of the sale, in that while it gained a lot of money for the state, it just created a private monopoly in place of a public monopoly. Up until then the British Gas Corporation had preferential rights to buy natural gas for domestic consumers. However, such arrangements were viewed as uncompetitive, and by the end of the decade, competition was applied to the natural gas sector.

Privatisation and liberalisation became the key ideologies of later Thatcherism. This was to underpin the next major target of privatisation in the energy sector: the electricity industry.

It was a mixture of privatisation and liberalisation which radically changed the energy sector. This led to various technological changes and what became known as the 'dash for gas' in the shape of many new combined cycle gas turbine (CCGT) power stations (CCGTs were then a new technology). At the time, the consensus saw this as a positive development, but with the hindsight of the energy crisis which started in 2022, this seems less certain. Higher gas prices (which began to rise from 2006) meant that electricity prices increased since the price of electricity from gas power plants sets the wholesale electricity price. Privatisation and liberalisation appeared to be a success in energy in the 1990s because of the availability of cheap natural gas supplies.

As discussed in Chapter 1, I describe neoliberalism as involving a new role for the state, whereby the state acts to set up and regulate markets that are deemed to be competitive. The aim is to achieve lower prices for consumers for goods and services: in this case energy. Under neoliberalism, the electricity and natural gas markets involve regulation to achieve (1) a set of basic criteria of quality (e.g. safe supply of energy services) but also (2) a carefully arranged market mechanism where effective competition can be assured. This involves a lot of regulations, and hence the term 'deregulation' is misleading. This is much different to nineteenth-century laissez-faire ideas, where the aim was to keep state regulation as low as possible so as not to interfere with the market. In this contemporary notion of neoliberalism, the state has the central role of defining the market and appointing a regulator.[7]

Advocates of liberalised markets argue that considerations such as environmental objectives can be incorporated into markets. This can be done through regulation, energy taxes and/or the creation of pseudo-markets, for instance markets in clean energy.

Neoliberalism in energy gives priority to what are in effect short-term price reduction objectives while offering no clear guarantees for energy security. This is clear from recent history, whereby the markets upon which the UK has relied for natural gas supplies have become dysfunctional and sent energy prices rocketing. Here of course, I focus on the idea of national energy security as meaning equitable access to reasonably priced and ecologically sustainable essential energy services for homes, industry and services. A further criticism of neoliberalism is that it has a bias towards means of delivering environmental and other objectives through artificially created markets, which create uncertainty and therefore increase, rather than reduce, costs for the consumer.

LIBERALISED MARKET ARRANGEMENTS

One crucial change in energy security arrangements was the ending of the British Gas Corporation's right of priority purchase of natural gas from the North Sea (for domestic consumers). The electricity industry was, until 1990, state owned and organised under the Central Electricity Generating Board (CEGB), but private companies could take advantage of natural gas from the North Sea. This opened the way for independently developed gas-fired power stations when the electricity industry was privatised and liberalised.

So, when the CEGB was privatised in 1990, the way forward was open for companies wishing to develop what was then the new technology of CCGTs. This happened because privatisation was implemented alongside market liberalisation, which opened energy markets to competition for a range of companies.

Previously, gas-fired power stations were gas-peaking plants (used on occasion), whose rate of conversion of gas into electricity was in the low 30s in percentage terms, or combined heat and power plants. In the latter case, the heat that was co-produced was used either in large-scale industrial contexts or in district heating systems. By the end of the 1980s, it became practical to run CCGTs with efficiencies approaching 50 per cent. These efficiencies were much higher than even the newest and largest coal-fired power plant. Since then, the CCGT efficiencies have increased beyond 50 per cent.

Upon privatisation, the CEGB was broken down into its generation, transmission and distribution elements. Initially, domestic supply was kept as part of the newly established 'regional electricity companies' (RECs) that ran the electricity distribution network. They could also part-own new projects for generating electricity.

The existing power station fleet was divided between three big companies, two being privatised and one holding nuclear power. Nuclear power was kept in the state sector before being sold to EDF in 2009. By then a number of electricity companies had emerged, owning various chunks of the electricity industry with holdings often combined with the (also privatised) gas industry. Under the old CEGB, the operation of power stations was determined by a 'merit order', under which the costliest power plant was reserved for when demand was highest.

After liberalisation, a competitive wholesale power market replaced the merit order, and operators were paid according to a market price ('marginal cost pricing'), as opposed to payments being made based (notionally, within the CEGB) on the assessed cost of electricity generation. RECs and other suppliers, and large consumers, bought their power needs on this new wholesale power market. This difference had crucial significance for the 2022 energy crisis in that all generators received the same sky-high prices rather than being recompensed for their (often much lower) costs.

The new market-based trading system meant that suppliers obtained the energy they needed to satisfy consumer demands from the generators, via the wholesale power market. Under the current system, generators sell power to suppliers through bilateral trades on a half-hourly basis. The generators have to provide the suppliers with the energy they have agreed to send them in advance, and in turn the suppliers must satisfy consumer demands for power. The National Grid operates a balancing mechanism which, in effect, irons out the small differences between the physical needs of the system to maintain stability and any shortcomings in the electricity trading system. It also makes sure that there is, on a day-to-day basis, enough generating capacity available to meet demands made by suppliers.

The 'dash for gas' started soon after the 1990 privatisation and market liberalisation. This happened after the RECs, along with independent companies, organised for a string of CCGTs to be built. This led, alongside the ending of restrictions on buying coal from abroad, to coal mine closures. It also led to a big increase in the consumption of natural gas from the North Sea.

The gas industry was liberalised in parallel with the electricity industry, with gas trading systems between generation and supply being established. Competition was allowed in the domestic retail sector from 1998. The commercial shape of the electricity and gas industries then resembled its present form, with restrictions on the ownership of different types of assets being removed. The system became dominated by half a dozen big energy companies which owned both electricity and gas assets, and which bought up the regional electricity distribution companies (the RECs) on an ad hoc basis.

The whole system was overseen by regulators, with the Office of Gas and Electricity Markets (OFGEM) created in 2000. OFGEM oversaw the electricity and gas industries. Wholesale power markets were organised based on disaggregated bilateral trading, with a balancing market run by electricity and gas system operators.

In this system, nobody seemed to have a clear responsibility for national energy security. The government had relieved itself of responsibility for energy provision and assumed that consumer interests would be served by the arm's length, regulated, competitive markets. Competition would ensure lowest prices. It was assumed, on the basis of ideological faith, that neoliberal arrangements would provide the best deal for consumers. Yet these arrangements failed when it came to energy security.

THE ENERGY SECURITY CONSEQUENCES OF NEOLIBERALISM

One consequence of neoliberalism has been great confusion about who is responsible for energy security, a confusion that has had enormous consequences for the energy crisis which emerged in 2022. As Tom Edwards, a consultant for Cornwall Insight, put it when talking about electricity supply, 'National Grid (ESO & transmission) has no obligation to keep the lights on, they are there to manage the system in a safe and efficient manner – Security of Supply is the markets/Government job'.[8]

In a 1995 White Paper, the government stated that:

[T]he Government has been working towards liberalisation of the energy sector. The Government is drawing back from direct involvement in the energy markets and will consider intervening only when there are compelling environmental, social or economic reasons to do so. The Government believes that, so far as possible, choices about primary energy sources should be left to market participants.[9]

GAS STORAGE, LIBERALISATION AND SECURITY

Yet if the market makes the choices, how is the market going to secure long-term security objectives for which there is no demand from the market? It may have been argued that greater natural gas energy storage was required to secure long-term British energy security objectives, but no action was taken by the British government to secure it. The market was unlikely to provide this aspect of energy security. The whole of the UK, as opposed to individual companies, was likely to make gains from long-term investments in energy storage, especially of the size needed to make a difference.

In fact, existing energy storage facilities were closed down. The largest gas storage facility – Rough, in the North Sea – was opened in 1985 during the nationalised era, but it closed in 2017 because the facility was proving uneconomic. Efforts were made to reopen the facility in late 2022, but even then UK gas storage amounted to only 2 per cent of annual natural gas consumption. This compares to the 20–25 per cent of annual consumption available in storage facilities in the case of other large western European states such as France, Germany and Italy.[10] Despite the increase in natural gas use in the electricity sector, which rose to account for around half of all electricity production, the UK was self-sufficient in natural gas until the turn of the century. However, the proportion declined thereafter, falling to around 40 per cent of UK gas supplied from the North Sea by 2021.

The central problem with the provision of gas storage, in a liberalised market system, was that individual gas supply companies would only consider the benefits of building gas storage in the context of their own interests in maximising revenue. This is very different from the national interest, which is in ensuring that gas prices are kept down at times of supply shortages. During the energy crisis of 2022, other European countries, with large amounts of gas storage, bought in gas supplies during the summer, when demand and international gas prices was relatively low. This enabled them to provide supplies at more reasonable prices during the winter, when global gas prices were much higher. But the UK was unable to do this because it had little gas storage capacity.

NUCLEAR POWER

The government did respond to some interest group lobbying about energy security, but this was mainly concerned with giving incentives to build and maintain conventional electricity generation capacities. One example of this was to boost the prospects for nuclear power. Nuclear power construction had been abandoned at the time of electricity privatisation, since none of the privatised companies believed that this would be profitable.

The new drive to build more nuclear power plants, which can be traced back to announcements by Tony Blair in 2005, was justified by the need to meet an expected rapid increase in electricity demand. Initially, it was hoped that the market would produce the outcome of more nuclear plants, as energy prices rose after 2005, led by oil price increases. However, little was done to enhance UK energy security, and this situation continues to this day, with no plants being commissioned.

This is despite a major incentive package for nuclear power being decided in 2010, and the first contract being signed for the Hinkley C nuclear power plant in 2013. Hinkley C's delays, and consequentially its construction costs, have been ever increasing. As discussed in Chapter 1 (to which we return in Chapter 4 in the section on France), modern Western industrial conditions do not facilitate the building of nuclear power plants to produce electricity. No amount of hopeful estimates will prevent the inevitable gross cost overruns and failure to deliver any electricity generation for many years. It therefore seems unlikely that the next proposed nuclear project, Sizewell C, could be built without a massive drain on the public purse or consumer energy bill. The costs of Hinkley C will be borne by the French government, which owns EDF. However, the French government will insist that the British government guarantees that British people will pay for the financial catastrophe of Sizewell C.

Certainly, if energy security was to be measured by affordability of electricity supplies, new nuclear power plants did not appear to provide it. Delivering their construction seemed to be a very long, drawn-out process. Politicians blamed each other for the failure, apparently oblivious to the problems with the technology itself.

CAPACITY MARKET AND PRICE CAP

In response to fears that there were no incentives, under neoliberal market arrangements, to ensure there was enough electricity generation capacity, a 'capacity market' was established in 2014. This gives incentives to conventional fossil and nuclear power plants, both existing and planned, to be available for generation if needed. The previous state-owned system, run by the CEGB, had been criticised for being costly. However, in order to buttress energy security in a neoliberal system that did not otherwise protect energy security, new, costly layers of incentives had to be added in the form of capacity payments to generators. These payments are made to all conventional power plants, although most would still be available without the capacity payments.

The capacity payments are added as a charge to electricity bills. Yet even these proved wanting when the energy crisis began to emerge in late 2021, because even though there might be, in the case of electricity, enough generating capacity, there was no mechanism to contain price rises. That is because of the 'marginal pricing' system of the liberalised energy market. This meant that all generators were paid the price of the highest (marginal) unit of supply.

Liberalised markets were championed because they emphasise competition as the means to restrain prices. But this only works where there is a surplus of supply over demand. However, perversely the market design ensured that energy prices were the highest they could be in 2022, when there was a supply shortage compared to demand.

There were also question marks about how much competition liberalised markets engendered. There has been some meaningful competition in the industrial and to a lesser extent in the commercial supply sectors. However, the retail electricity supply market has been handicapped by poor conditions for competition. It is very expensive for suppliers to market their services to many millions of domestic consumers. Many consumers, themselves short of time, do not bother to 'compete' for the best deals. Suppliers resorted to short cuts, rewarding those who switched suppliers, whereas most consumers, who did not spend time trying to switch suppliers, paid a premium.

A further problem with the liberalised arrangements is that the supply market will not, on its own, deliver the sorts of balancing techniques (of supply with demand) that will be increasingly needed as the proportion of fluctuating renewables grows bigger. It is not possible to prevent suppliers from gaming their prices upward or downward in this so-called competitive market. In recognition of this, the price cap was introduced in 2019 to deal with this problem. However, this challenges the reasoning for having a competitive market in the first place if a government agency, OFGEM, was setting prices that would be no less advantageous than the allegedly competitive fixed-term deals.

Yet the supplier's price forms only a small part of the total consumer bill. Wholesale energy costs are at least three to four times the level of the supplier (operational) costs.[11] It is the high wholesale prices that caused the energy price crisis which began in 2022, and there is nothing that suppliers could do about that.

HOW NEOLIBERALISM FAILED RENEWABLE ENERGY

A further problem with the liberalised market was that it did not do enough to promote innovation though renewable energy technologies. The neoliberal bias meant that incentives to promote renewables had to be offered using methods that looked like the market was setting the prices. When the Labour government sought to meet ambitious targets for renewable energy production, it established the 'Renewables Obligation' (RO) (in 2002). This set energy suppliers an obligation to supply ongoing increases in the proportion of their supply through renewable energy. This involved a market in renewable obligation certificates (ROCs). Renewable generators were given an ROC for every MWh of production which could be sold to suppliers to meet their obligations to supply renewable energy.

Unfortunately, the RO proved to be expensive. Uncertainty over future ROC values increased the cost of borrowing money for the projects. Also, the lack of sufficient projects to meet the RO target increased the cost of the RO scheme since the value of the ROCs increased. The system was eventually changed to a 'contracts for difference' (CfD) system. This means that there is a fixed (inflation-proofed) price paid to generators for units of renewable energy produced for 15 years. However, more than a quarter of UK electricity generation was incentivised by ROCs when the 2022 energy crisis struck. There was no cap on the wholesale power prices earned by the renewable generators under the RO. Hence the renewable generators were earning windfall profits.

British renewables have expanded substantially. About 29 per cent of UK electricity came from wind and solar power in 2022, nearly 25 per cent from wind power alone.[12] However, its delivery has been more expensive than it should have been, because a lot of the incentives were based on neoliberal ideology. Planning restrictions and a lack of reliable incentives for renewable energy developers have also slowed progress.

Following the 2010 general election, efforts to develop an energy efficiency programme for buildings were wound down. The 'Carbon Energy Reduction Target' (CERT) scheme, which had been started by the Labour government in 2008, was ended in 2012. The steady decline in government funding for insulation measures can be seen in Figure 2.3. The number of cavity wall, loft and solid wall insulations carried out in the 2018–22 period was only 7 per cent of the number carried out in the 2008–12 period. This meant that millions of consumers had much higher bills because of a lack of insulation.

The feed-in tariff programme for small renewables (mainly rooftop solar PV), established in 2008, ran for a few years. The

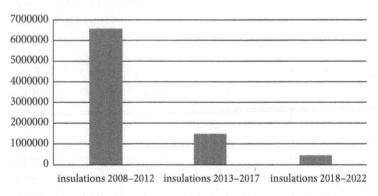

Figure 2.3   UK government-funded cavity wall, loft and solid wall insulation measures, 2008–22

Sources: OFGEM CERT Final Report 2013, UKGOV, www.gov.uk/government/statistics/household-energy-efficiency-statistics-headline-release-august-2023, House of Commons Library: Community Energy Saving Programme.

financial incentives for homeowners to deploy solar panels were then reduced, before the scheme was finally abolished in 2019 leaving no practical incentives to develop solar PV on homes. A couple of electricity suppliers (in particular Octopus Energy) have been paying high prices for solar exports, but the majority pay very little. It is difficult for community solar or wind farms to set up in business since they are not able to compete for fixed price contracts under the government's auction scheme for renewable energy developers.

Indeed, renewable energy deployment slowed in the UK after the policies of a majority Conservative government came into effect. In the earlier years of Conservative rule, in 2010, the government was held back from doing more damage to the renewables programme by the presence of Liberal Democrat energy ministers in the government. But since their departure after the 2015 general election things slowed, especially with regard to deployment of onshore wind and solar PV. This can be seen in Figure 2.4. There was a great deal of Conservative pressure against the development of onshore solar and wind farms. This has contributed to a below par increase in renewable energy generation in the UK compared to the rest of the world. While renewable energy generation worldwide increased by nearly 70 per cent in the 2018–22 period, in the UK it increased by just under 23 per cent.[13] The failure to install more cheap renewables cost the country a lot of money when it came to the energy crisis, starting in 2021.

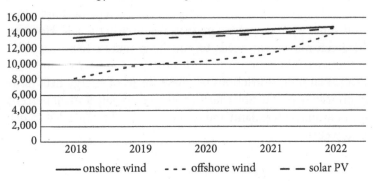

Figure 2.4   UK capacity of onshore wind, offshore wind and solar PV, 2018–22, in MW

Source: UK Digest of UK Energy Statistics 2023 published by UK government.

Indeed, the UK has (since 2019) fewer incentives for residential solar PV than even most US states. At the beginning of 2023 there were no incentives for solar panels on homeowners' roofs. Moreover, most of the energy suppliers paid very little for exports of solar electricity to the grid.

We can see that neoliberal arrangements failed to provide energy security during the energy crisis. That crisis was made much worse by the nature of liberalised markets themselves. When it comes to pursuing the net zero target for greenhouse gas emissions (agreed by the government in 2019), the interests of the big energy corporations are given priority over and above renewable energy.

A prime example of this occurred in March 2023 when the government announced a multi-billion-pound programme to fund CCS and SMRs.[14] It stretches credulity to believe that after decades witnessing the nuclear industry gradually scaling up its designs for nuclear power plants, now the answer to nuclear's problems is to make the reactors small. CCS is a technology that will protect the corporate fossil fuel industry, using an expensive technology that at best collects no more than 90 per cent of carbon emissions, which is not high enough if the objective is to reach zero carbon emissions.

Then Prime Minister Sunak announced a slowdown of net zero policies, as if they were fast anyway! This focused on relaxing targets for banning non-fossil fuel vehicles and fossil fuel boilers, postponing measures to ensure better insulation of energy-inefficient rental properties and threats to introduce planning restrictions on solar farms. This is on top of the existing effective ban on onshore wind.

Offshore wind power is a massive resource in the UK. However, the green movement needs to put much more effort than it has in the past into supporting both onshore wind farms and solar farms. Planning restrictions on these forms of energy need to be vigorously opposed. Solar farms will take up only a small percentage of available farmland and do not need to be restricted to help food security.

THE 2022 ENERGY CRISIS

It is now routinely stated that the natural gas price crisis that has affected the UK, and Europe in general, was the result of the

Box 2.1 Have UK decarbonisation policies really been effective?

The UK's progress in cutting carbon emissions by almost half between 1990 and 2022 seems relatively impressive. But a lot of this has little or nothing to do with deliberate climate or environmental policies. Britain is in a respectable but not leading position in terms of lower greenhouse gas emissions. Among 38 Organisation for Economic Co-operation and Development (OECD) countries, the UK is eighth lowest in terms of greenhouse gas emissions per unit of gross domestic product (GDP) (Switzerland is the best) and eleventh lowest in terms of per person emissions (Costa Rica comes top on this).[15] Yet even these positions may not have much to do with the UK's climate policies per se.

A lot of the UK's reduction in carbon emissions is concerned with a shift from using coal to using natural gas, mainly in electricity production. This change came after privatisation and liberalisation policies took effect in the 1990s (although the inspiration for this had little to do with climate policy). The switch to gas took precedence over innovating with renewable energy. This left other countries such as Denmark, Germany and China to dominate the production of wind and solar power. It also reduced the energy security of the UK by making it more reliant on imports of natural gas.

Yet the UK does not do that well on specific green energy policies. The UK's renewable energy programme has produced rather less wind and solar power per person compared to most other Western states.[16] The UK's housing stock is regarded as having very poor energy efficiency levels compared to other European countries.[17] The UK competes with Hungary for the lowest per capita installation rates for heat pumps in Europe.[18]

Besides a switch from coal to gas (which has also been happening to a greater or lesser extent in other Western states), what has contributed to falling UK carbon emissions? A big contribution to depressing UK carbon emissions comes from structural economic factors. The UK is deindustrialised compared to other major Western economies, and so will consume a lot less energy to produce its economic outputs.[19] Also, the UK has rather lower floorspace per person compared to many other countries,[20] and so its buildings consume less energy (something which counteracts the relative energy inefficiency of UK buildings). These factors are unplanned consequences of British policies and certainly not the outcomes of deliberate climate policies.

In fact, the UK would be much lower in the OECD rankings for positive outcomes of climate policy if it were not for these factors. The UK certainly fails as a climate 'leader' in those terms. To put it another way, if the UK did have policies for renewable energy that are as good as, say, Denmark, then the UK, with the same structural economic factors as it now has, would be doing much better in leading the OECD countries in having lower emissions rather than merely coming in eighth or eleventh in the rankings.

Russian invasion of Ukraine. However, that was only part of the problem. Analysis of the data reveals that from May 2021 onwards, natural gas prices in Europe were spiking at three times their earlier (twenty-first-century) levels – that is, well before Russia invaded Ukraine on 24 February 2022.[21] The fact that domestic UK consumers only felt the increases in price levels later in 2022 is an artefact of the delay produced by the regulatory price cap, which compelled suppliers to delay price increases.

The spiking price of European natural gas was a consequence of increasing demand on global markets. There has been a big increase in the demand for natural gas from countries in the Far East, including South Korea, with demand from China doubling in five years. This undermined the reliance of the UK on global supplies of natural gas, shipped as LNG. The problem with oil and gas resources is that their supply is finite at any one time and subject to imbalances with demand. Price spikes occur regularly, and when they do the oil and gas companies make tremendous profits at the expense of struggling consumers.

This crisis was intensified in the rebound from the Covid slowdown. As with a range of other commodities, supply was slow to respond to increased post-Covid demand. The world LNG supply is relatively much smaller than the supply of oil on ships. Shortages of piped supplies can easily be offset by shipping in. Prices on the world LNG market increased greatly. This meant that Russia's Gazprom was in a good position to gain a great deal more income by small reductions in supply, requiring European consumers to pay much higher prices. Then the Ukraine war started, and European countries sought to reduce Russian consumption in order to counter the weaponisation of gas supplies by the Russian government.

Neoliberal policies exacerbated this problem. European countries have, in the last decade, relied increasingly on spot market natural gas prices, rather than long-term contracts, for natural gas, believing that this produced lower prices. However, this policy offered no long-term energy security, something which became clear as 2021 wore on. The UK was the keenest advocate of spot market trading and had no substantial long-term contracts.

Even worse for the UK, it had practically no gas storage capabilities compared to countries such as Germany, France, Italy and the Netherlands, who had 20–25 per cent storage capacity compared to annual consumption. Large storage capacities meant that these countries could buy natural gas in the months of the year when heating demand was low and when gas supply prices were thus relatively lower. These supplies could be stored for use in the winter when natural gas prices on the global LNG market would be extremely high. What made it even worse for the UK was that the country was reliant on natural gas for its energy consumption in houses that are, on average, much less energy efficient than in the case of other European countries.

Liberalised energy markets compounded the problems in the electricity sector. In theory, much electricity generation was bound up with new renewable and old nuclear power plants that were now much cheaper than electricity from gas-fired power plants, the nuclear power plants having had their capital costs paid for by the state many years ago. However, all electricity generation was paid at the same price as the most expensive source. This meant that nuclear power and renewable energy power plants were paid the same very high prices for their power. The liberalised wholesale power market was supposed to reduce prices through competition, but now its marginal pricing mechanism was actively increasing these prices, to the gross detriment of consumers.

The energy price rises caused havoc in the energy supply market, with several suppliers going bankrupt. They misjudged their 'hedging' positions by assuming that future prices would not rise as they did. Consumers had to suffer increased regulatory costs of around £2.7 billion to cover for these losses.[22] Yet such losses were small beer compared to the cost of shielding consumers from the immense price rises, that fed through to wholesale energy costs.

A further anomaly arose in the renewable energy sector where a large proportion of projects were still being given money for ROCs, although they were receiving power prices two or three times their normal level. This further highlighted the inefficiency of the market-based renewable energy obligation (the government tried to persuade the developers to move on to fixed price CfDs without much apparent success). In the end they levied the developers with a 'windfall profits' tax. Windfall profits were also levied on oil and gas companies, although these companies could offset investments in new oil and gas fields against their liability, something not allowed in the case of the renewable energy companies.

All in all, the UK, which had blazed the trail in establishing liberalised energy markets in the 1990s, was very badly affected. The government introduced a price cap and borrowed billions of pounds to support it, which added to the already large government debt levels.

A big question is whether this calamitous situation could have been significantly alleviated if, since the 1980s, the UK had followed a different path with regard to the state relationship with the energy sector. In short, could and would things have been much better if energy had remained at least mainly state owned, as it had been at the beginning of the 1980s?

WHAT IF THE ENERGY SECTOR HAD NOT BEEN PRIVATISED AND LIBERALISED?

Consideration of what would have happened if energy had not been privatised requires a counterfactual explanation. This inevitably involves speculation, but some plausible assumptions can be made, and some comparisons with other countries can also be useful.

We start with oil and gas exploitation. If BP and the British National Oil Corporation had not been sold off when Margaret Thatcher's government took power, then tax receipts from oil and gas extraction would have been larger. In 2022, oil and gas revenues rose to 65 per cent of profits with the imposition of 'windfall' taxes. For much of the period, effective tax rates have been much lower than this. For nationalised assets of course, the state would have taken 100 per cent of the profits, which is a lot better. There is also a likelihood that a publicly owned natural gas industry would have had long-term contracts to supply the domestic gas supply industry. That was certainly the condition in force before privatisation. If so, then the energy price crisis in the UK in 2022 would have been substantially eased. As it is, the privately owned oil and gas companies operating in the North Sea sell the natural gas at prices governed by ultra-high world gas market prices.

Turing to the electricity supply industry, we have to consider what would have been different if the industry had remained in state hands. An economist's analysis of the first few years of privatisation/liberalisation concluded that while labour costs had declined, this was counterbalanced by increases in capital inputs:

> The results of this paper suggest that the productivity record of the UK ESI [electricity supply industry] after privatisation can be summarised as being unremarkable. Although labour productivity growth rates were very high in the period since privatisation, this was more than counterbalanced by increases in other inputs. Relative to its own past experience, or that in other countries, productivity in the UK privatised industry does not appear to have shown any pronounced improvement.[23]

If Thomas Picketty's analysis is used then one could argue that a decline in the number of employees was counterbalanced by a greater return to capital and thus wealth accumulation.[24]

Electricity prices may have been modestly lower because of the introduction of CCGT technology, as a consequence of greater competition. However, the increasing reliance on gas-fired power stations may have undermined energy security in later years for two reasons. First, because from the time of the 2022 energy crisis, gas-fired power plants produced very expensive electricity. Second, because the turn to using large quantities of North Sea gas, to be burned in the CCGTs, exhausted natural gas reserves in the North Sea at a much more rapid rate than would otherwise have been the case.

On the other hand, the replacement of coal-generated electricity by gas-fired generation did considerably reduce carbon emissions. This is partly because natural gas has a lower carbon output per energy unit than coal, and partly because the CCGTs are more energy efficient generators compared to coal-fired plants. Of course, CCGTs were not built because of any climate change policies – in the 1990s climate policy was in its infancy and not the central driver it is, or at least supposed to be, today.

In speculating about what the electricity generation choices might have been if the CEGB had not been privatised, it is plausible to say that there might have been one or possibly two more power stations built the same size and nature of Sizewell B. Certainly the earlier version of Hinkley C was given planning consent at the time of privatisation, but nuclear building plans were scuppered because they were uneconomic when faced with a competitive electricity market. So, the relatively greater carbon emissions that would have accompanied more coal-fired plants remaining in production might have been largely offset by the fact that (1) more nuclear power (which produce low-carbon energy) would have come online and (2) the fact that some CCGTs would have been commissioned by the CEGB to replace the higher-carbon-generating coal-fired power plants. Given the decline in CCGT power costs in the 1990s, they would likely have been favoured rather than building many nuclear power or coal-fired plants.

Although the major conventional power plants would have remained in public ownership, it is doubtful that renewable energy development would have been left solely in the hands of the CEGB. That conclusion is reached by simple international comparisons. Where innovation is concerned, apart from conventional power plants, it always seems that substantial programmes in renewable development – certainly for wind and solar power – have been organised on a competitive basis. This includes the various cases where renewable energy has been organised in the context of publicly owned energy systems. In fact, these are quite common in Europe. In the cases of Norway, Sweden and Denmark, major energy companies are state owned. Statkraft (electricity) and Equinor (oil and gas and now some renewables) in Norway, Vattenfall (electricity) in Sweden and Orsted (formerly Danish Oil and Natural Gas (DONG)) in Denmark. But their own renewable energy markets are competitive – they do not give these companies a monopoly over renewable energy. Indeed, these companies compete for renewable development in other countries. This includes the UK, where over 40 per cent of the project equity in offshore wind farms is owned by foreign state-owned companies.[25]

Certainly, the offshore wind programme in the UK has been a considerable success. However, it should be noted that this is because of the high level of state intervention rather than the lack of it. Offshore permitting processes were streamlined, with most planning disputes for offshore wind paradoxically occurring because of the onshore permissions needed. Nevertheless, when planning problems have occurred, the government has allowed contract extensions to keep the projects going. Financing has been well organised with very high contract prices offered for the earlier projects. However, since 2017 the price, for agreed contracts, has tumbled, with the exception of 2023 where the increase in interest rates pushing up capital costs meant that offshore wind became too expensive for the Treasury's liking. No offshore wind contracts were issued as a result.

In short, if UK electricity had not been privatised, then the publicly owned CEGB (or some quango such as the current Low Carbon Contracts Company) would most likely have had to issue generation contracts for renewable energy through competitive tendering. But the energy would, one assumes, have been sold under fixed price contracts to the CEGB. Probably there would be a slice of that won by a British publicly owned generation company (a company associated with the CEGB), as well as foreign publicly owned companies. A key difference with the current arrangements would be that the electricity generation would have been bought and controlled by the publicly owned company (CEGB). They would not have had to pay sky-high prices to private companies via the wholesale power market for renewable energy power supplies that ought to have been a lot cheaper.

Of course, concluding that our energy situation would be much better if the energy situation had not been privatised and liberalised is one thing, but it is too late to curb much of the present crisis through nationalisation. This is because compensation would have to be paid at market rates – there are too many international treaties and lawyers to avoid that. But there is a strong case for selective public ownership to look after future national energy security. However, the domestic retail supply part of the energy industry (which is really a natural monopoly) will not cost much to put back into public ownership, and there are good reasons to do this. Moreover, a new state company could be created to compete as a renewable energy developer.

What is the case is that greater state planning and direct public ownership is needed if energy security interests are to be protected and the task of reaching net zero greenhouse gas emissions by 2050 is to be achieved. For example, much greater amounts of energy storage need to be built for two reasons. First, this is needed to protect national energy security. Second, this is necessary to provide long-term energy storage to balance renewable energy, as the UK approaches deriving 100 per cent of its energy from renewable energy sources. These two objectives can be combined, since renewable electricity from wind and solar can be used to produce carbon neutral methane.[26] Yet the market is not going to provide the incentives necessary for this. Only direct state intervention can achieve it, preferably through actual ownership of assets.

It should be added here that although the EU generally has suffered through the adoption of a neoliberal political economy (substantially influenced by the UK along the way), leaving the EU brings energy disadvantages for the UK (see Box 2.2). The fact that France and Germany increased state ownership of energy assets in 2022–3 (see Chapter 4) implies that EU membership does not prevent public ownership. Countries such as Sweden,

---

Box 2.2   How UK energy has fallen behind others in Europe since Brexit

First, there has been an increase in electricity prices – The UK left the European Single Market in 2020, meaning that electricity trading between the UK and the rest of Europe has become more complex. Instead of selling to one digitalised market, the UK trade in electricity through interconnectors is now done in a series of pieces to different markets. According to the consultants Barings, as covered in the *Financial Times*, this cost the UK up to £250 million in 2021 and £440 million in 2022.[27]

Second, the EU target for solar PV is 600 GW by 2030 while the UK government is only targeting 70 GW by 2035 (much less pro-rata based on population than even the 2030 EU target).

Third, the EU's renewable energy target is for 50 per cent of all energy to come from renewables by 2030, whereas in the UK the target is only for all electricity to be decarbonised by 2035, which will be much less than 50 per cent of all energy.

Denmark, France and Norway (which is in the single market only) have large amounts of state ownership of energy. However, there are EU regulations insisting that there is some competition in the energy supply sector. This is odd given that the domestic supply sector, as argued here, does not involve the possibility of useful competition.

WHY ENERGY SUPPLY SHOULD BE PUBLICLY OWNED

Energy supply (as distinct from generation or even transmission and distribution) should be brought into public ownership – but not (just) for the reasons that are often given. In 2022 the British TUC union published a plan for taking the major energy suppliers into public ownership, although confusingly not all of the suppliers. They included proposals for 'a social tariff capped at 5% of income for low-income households' and to 'restructure tariffs to provide all households with an initial free energy allowance and increase the cost per unit for high-consumption households'.[28] The proposal would cost less than £3 billion.

The TUC proposal only talks about energy suppliers and not generation or transmission/distribution. Nationalising all of that would be far more expensive, costing around £90 billion according to the UNITE union.[29] Energy supply is the part of the industry that is best known (and blamed by the public for high energy prices). Perhaps the simplest and most effective design would be to make domestic energy supply a publicly owned monopoly. Public ownership of the domestic retail energy supply could involve taking on existing contracts with generators. Competition for supplying industrial energy consumers could be retained using a wholesale electricity market.

The sort of competition among generators that is important is that which has produced low prices for the supply of renewable energy from wind farms and solar PV farms using CfDs. The energy regulator, OFGEM, has ensured that these low prices are passed through to consumers by taking these lower-priced sources into account when calculating a price cap that the suppliers can charge. Public ownership of the energy supply would replace the price cap in ensuring that the savings to consumers from cheap renewable energy projects were passed on directly to consumers.

The point I want to make here is that competition in the energy supply sector has failed – certainly in the retail sector – and is unlikely to serve the consumer well, whatever tinkering is done. Bringing it into public ownership and delivering the supply on a public monopoly basis (for domestic consumers) makes sense for three very good reasons.

First, this would constitute a monopsony, where the buyer has a monopoly, at least for domestic energy consumers, although larger energy consumers could contract separately in a continued wholesale energy market. In terms of economic theory, monopsonies (and oligopsonies) have an incentive to lower costs by squeezing the product manufacturers (the generators in the case of energy). Having a monopsony works to the advantage of British consumers in the case of the National Health Service.[30] The single supplier can do so through greater bargaining power, and can also reduce unit costs by reducing energy purchases and avoiding the most expensive options. That being said, an increasing volume of generation used by the public supplier would consist of CfDs issued by the government through the competitive auctions it now organises annually. All energy suppliers, whether under a publicly owned or privately owned system, should be given targets and mandates for the deployment of insulation and heat pumps in existing buildings.

Second, the savings from cheap renewable energy schemes given a CfD by the government would pass straight through to the consumer under a publicly owned energy supply arrangement. A frequently mentioned solution to the problem of expensive gas power setting the price for all wholesale energy is to separate out markets for fossil and non-fossil generation. This has the downside of increasing the complexity of an already very complicated system. However, perhaps an even bigger problem is that electricity markets are being increasingly internationalised, and this in itself could lead to electricity spikes on the markets for renewable energy. The system using CfDs with renewable energy generators can simply be combined with price controls operated by a publicly accountable supply system.

Third, when it comes to balancing increasing supplies of fluctuating renewable energy, public ownership of the supply sector could also do a much better job of advancing arrangements for demand-side management (DSM). DSM is a technique that can be used to balance fluctuating renewables, on a daily basis, by incentivising consumers to consume at one time rather than another (as discussed in Chapter 1). Yet efforts to set up basic conditions to allow this in the UK have so far stalled. The regulators, OFGEM, have said that 'half-hour statements' (which allows electricity trading at the retail level) will not be in place until 2026. On top of this, the government seems set to allow DSM on a 'market' basis. In other words, only some electricity suppliers are likely to offer the service on the (very limited) basis of what might benefit the supply company rather than the electricity system as a whole.

A problem is that, just as the liberalised energy system failed to provide enough gas storage to serve the UK's national interest, so a liberalised electricity supply system will fail to provide the amount of balancing services that the national electricity system needs. It is unlikely that all electricity suppliers will offer effective incentives to shift electricity usage from one time to another. Consumers will need to be given incentives to shift their demand and to install equipment such as home-based batteries. This is a cost to the system which needs to be shared. Individual competing suppliers will be unwilling to incur extra costs to benefit the system in general.

There are some good arguments that energy transmission and distribution (which are natural monopolies) should be publicly owned. Set against this is the argument that the £30 billion or so that this would cost (according to UNITE) might be better spent on the urgent task of upgrading the networks to accommodate more renewable energy capacity.

COMPETITION THAT WORKS

Competition can work to the interests of the consumer and the environment in the generation sector. However, the government (via an agency) should issue long-term power purchase agreements (PPAs) through a competitive system. This competitive system would involve both public and private companies. In fact, a lot of the 'competitors' for renewable energy contracts are (already) publicly owned companies from Scandinavia. There should also be a publicly owned renewable energy company established in the UK to compete with the other private and public companies for renewable energy (and energy storage) contracts.

The sort of competition among generators that is important is that which has produced low prices for the supply of renewable energy from wind farms and solar PV farms. This has been achieved using CfDs, which pay a guaranteed price for the renewable energy that is generated for 15–20 years. If the price of electricity on the wholesale power market is more than this guaranteed price, the operators pay money back to the consumer, and vice versa if the wholesale price is lower. This helps to protect the consumer from profiteering during times of high marginal wholesale power prices. The energy regulator, OFGEM, has ensured that these low prices are passed through to consumers by taking these lower-priced sources into account when calculating a price cap that the suppliers can charge consumers.

Competition among prospective generators occurs when the government invites prospective generators to make tenders to supply the renewable energy at a price that they can offer in order to make a profit. The government holds reverse auctions in which the developers bidding the lowest price for payments for each MWh generated win the contracts to supply an amount of capacity that the government wants. The CfD system has replaced (for new projects) the previous (and expensive) RO system of incentivising renewables. The CfDs came about because the government realised, in 2011, that nuclear power could not be financed through the so-called market incentive system used with renewable energy, and so a fixed price for electricity produced had to be guaranteed. Ironically this has been abandoned for nuclear (after the experience with contracting Hinkley C) and the government is instead facing the prospect of writing a blank cheque (on behalf of UK energy consumers) to build Sizewell C.

A GREAT BRITISH ENERGY COMPANY?

In September 2022, the Labour Party proposed that a publicly owned energy generator, a 'Great British Energy Company', be established to invest in non-fossil fuel energy generation. This represents a potentially valuable and relatively low-cost way of increasing public ownership in renewable energy, a rapidly expanding area, if the company was restricted to renewable energy and energy storage rather than nuclear power.

In September 2022, Labour announced that:

> [T]he role of GB energy will be to provide additional capacity, alongside the rapidly expanding private sector, to establish the UK as a clean energy superpower and guarantee long-term energy security. ... A publicly-owned company is the best way to ensure that the British people can derive the benefits from the power that we create on our own shores – delivering cheaper bills, good local jobs, and bringing money back into the public purse.[31]

However, the mission statement is too vague, implying that it could be a vehicle for attempts to develop nuclear power. Nuclear power is a notorious loss-maker. No nuclear power project has, at least this century, been brought in anywhere near on time and on budget in the West, and always at a very high price compared to contemporary new wind and solar power schemes. There is a danger that the proposal will sink if it is associated with nuclear power.

On the other hand, in October 2022 the Welsh government announced plans to form a renewable energy company in Wales. Learning from the experience of projects in Wales owned by the Swedish state company Vattenfall, the Welsh state company will initially focus on developing wind farms on land owned by the government. The profits will fund an energy efficiency roll-out in Wales.[32] Prior to this, the Scottish pressure group the Common Weal has advocated action by the Scottish government to establish a publicly owned energy company which would help communities develop energy assets.[33]

Despite the widespread bemoaning of the lack of natural gas storage facilities in the UK, there has been surprisingly little call for a publicly owned facility to engineer substantial new gas storage facilities. There is no shortage of possibilities for this in salt aquifers in the North Sea. Such facilities can be used as a bridge between the current and medium-term need for natural gas storage, and the future need to store gas derived from renewable energy to support an energy system based on renewable energy. Renewable electricity can be turned into either hydrogen or synthetic fuels such as methane (using processes involving electrolysis) and stored over the long term to cater for years which have lower wind production than average.

UK GREEN ENERGY: THE NEED FOR RADICAL STATE INTERVENTION

The Conservative government committed to the net zero target in 2019 and it is true that UK carbon emissions have fallen substantially since 1990, but much of this has little to do with policy on green energy. As detailed in Box 2.1, the substantial decline in UK carbon emissions since 1990 has more to do with structural aspects of the British economy. Much of the planning and original momentum for the renewable energy programme was established towards the end of the term of the last Labour government and was partially kept in place until 2015 by Liberal Democrat energy secretaries.

However, in 2023 (at the time of writing) investment in renewable energy was falling and the government has maintained only a minimalist energy efficiency programme. Onshore wind has been effectively banned in England since 2015. Only modest programmes exist to promote heat pumps. Even banning fossil fuels in new buildings is taking much longer than needed. The government did set some targets, for example that solely fossil-fuelled motor vehicles would not be sold from 2030, but delivering the policies needed to achieve the targets has been lacking and targets for improving energy efficiency in rented buildings are being delayed. The targets for both solar PV and wind power are far lower than that needed to fulfil the electrification of the UK economy.

There are now, in effect, two price levels for electricity in the UK's power system. One is the level of the price paid to renewable generators under their CfDs, whose prices have been set by competitive auctions. The second is the (typically higher) price that will be paid to gas power station generators (in future using hydrogen or gas derived from renewables). An interventionist system (preferably involving public ownership of retail energy supply) is needed to ensure that the consumer pays for the schemes on CfDs at the cost of paying the prices set out in those contracts, not at the higher 'market price' set by companies producing gas for balancing.

The Scottish government's approach has been nearer what is needed given their much larger quantity of potential offshore leases compared to what is happening in England. Onshore wind continues to have planning support in Scotland. The Scottish government implemented building regulations favouring heat pumps well before the lazy timetable followed by Westminster for England.

The neoliberal approach of leaving targets to be achieved by market means has consistently failed to achieve targets in a cost-effective manner for the consumer. The main successes (e.g. fixed price contracts for renewable energy) have come about after UK governments found that more market-based systems did not work very well. The energy system is still based on neoliberal regulation that does not give sufficient priority to energy security and climate targets. OFGEM, seen as the 'referee' for assuring even-handed competition, is not suited to organise the energy transition.

A team of academics, including Rebecca Willis and Catherine Mitchell at the University of Exeter,[34] have proposed that: (1) an Energy Transformation Commission should be established to help steer the energy transition; (2) responsibility for regulating codes that run the electricity market should be given to a combined System Operator; (3) regional distribution energy service agencies should replace the current distribution network operators and have responsibilities for increasing demand-side flexibility and balancing within their areas; and (4) local authorities should take on responsibility for planning for the low-carbon transition.

In addition, a series of other interventionist measures need to be taken to boost renewable energy. These include increasing the capacity of CfDs put up for auction from onshore wind, solar farms, offshore wind, tidal stream, wave energy and geothermal, at appropriate prices for each technology. Rapid investment in training of personnel and investment in ports is urgently needed to achieve the government's target of 50 GW of offshore wind by 2030. Community renewables groups should be offered long-term PPAs for at least the same prices as received by large-scale solar or wind farms. There needs to be increased incentives for rooftop solar PV, and planning restrictions on onshore wind and solar should be removed. A publicly owned renewable energy generation company should be established with substantial seed corn funding. A state-owned company to develop long-term energy storage facilities should be established.

A series of measures needs to be taken in buildings. Fossil fuel boilers should be immediately banned from new buildings, solar PV and batteries on new buildings need to be mandatory and the fabric efficiency of new buildings needs to be improved. There needs to be a large and growing roll-out of heat pumps in existing buildings by giving energy companies an obligation to install increasing numbers of heat pumps.

Local councils need to be given a statutory duty to make plans to build heat networks served by large-scale heat pumps and local authorities need to be given access to sufficient capital funds for such purposes. This should release billions of pounds annually for the construction of these networks rather than the relatively small amounts of funding currently available under government schemes. A major buildings energy efficiency programme similar to that started in the last years of the Labour government needs to be started quickly. The domestic retail energy supply should be brought into public ownership so that DSM can be effectively organised and incentivised.

Most of these measures will be cost-neutral to the public finances or ones that save consumers money. Electricity levies needed to fund any of the measures will be smaller than currently planned since the building of new nuclear power stations (including Sizewell C) would be cancelled.

CONCLUSION

The story of British energy policy since the 1980s has featured the rise of neoliberalism and also, according to the analysis in this chapter, its failure to protect energy security. The government pursued a hands-off approach to the use of resources and technology choices in assuming that private competition, in specially designed energy markets, would produce the lowest prices for consumers. There was a big shift away from coal burning towards using natural gas, which has reduced carbon emissions, but then the government slowed carbon emission reductions by protecting natural gas producers. A strategy of privileging natural gas over renewables became pronounced in the years after 2015. Renewable energy and energy efficiency policies were cut back in favour of what has turned out to be disastrously expensive natural gas. The carbon reductions that have occurred have been the result of accident rather than design (see Box 2.1).

There were reductions in electricity prices during the 1990s which were at least partly due to reductions in global energy prices during this period. This big increase in electricity generation using natural gas rapidly depleted the limited gas resources in the UK. This was associated with the end of the preference given to domestic gas consumers in the supply of natural gas from the North Sea. Hence UK consumers were more exposed to volatile global markets for LNG. The liberalised energy market did not lead to an increase in natural gas storage to compensate for the exposure to potential price volatility in international gas markets.

Oil and gas assets were sold off to private interests, resulting not only in a loss of revenues from oil and gas production (both in the UK and abroad) but also a lack of preferential contracts to supply UK consumers with natural gas. The natural gas and electricity retail markets failed to function effectively and racked up costs for consumers to pay for failures in the liberalised market arrangements. Meanwhile multinational energy corporations have made huge profits during the energy crisis, starting in 2021, with costs that have been passed onto energy consumers. Much of this could have been avoided if energy had not been privatised and liberalised.

Hence, Neil Kinnock, quoted at the beginning of this chapter warning of the dangers of becoming reliant on imported natural gas and coal, was being prophetic. However, rather than relying on homegrown coal from the Welsh mining communities which he used to represent, the future lies in harnessing renewable energy sources, including the wind, sunshine, tides and waves. It also lies in reducing energy demand through electrification and other methods.

There was a slow realisation that technological change, through promoting renewable energy, was necessary. However, when the major programme, the RO, was launched, it was organised through an artificial market in incentives. Although it was supposed to reduce costs, this made the renewables programme more expensive than has been organised elsewhere through feed-in tariffs. When the energy crisis struck in 2022, renewable energy promoted under this scheme was paid the same sky-high rates for electricity generation as gas-fired power stations, even though the costs were much lower.

The way out of both the energy crisis and the climate crisis is based on renewable energy and energy efficiency. This needs to go hand in hand with strong state intervention in the energy sector. Pragmatic, practical considerations dictate a series of public ownership measures. This includes public ownership of the retail energy supply and the establishment of a state-owned renewable energy development company and a state-owned company for long-term energy storage. In addition, local authorities need to be able to develop local authority-owned heat networks served by large-scale heat pumps. Where there have been successes in renewable energy policy, such as in offshore wind, this has been because of well-organised state interventions rather than laissez-faire approaches.

Interventionist measures are essential if the cost advantages of cheaper renewable energy schemes are to be passed on to the consumer. The issue of CfDs to renewable energy generators has enabled the consumer to benefit from the cost savings of cheap renewable energy projects. These gains will be protected from leakage to profiteering companies by bringing retail energy supply into public ownership.

A crucial additional reason for public ownership of the domestic energy supply is that balancing the electricity market (which is needed as more renewable energy is used) may not be done effectively if it is left up to allegedly competing energy suppliers. As with the case of the failure to build enough gas storage, non-market objectives can only be achieved through direct government intervention.

The problem of energy price volatility, as well as greenhouse gas emissions, will also be reduced by the speedy adoption of heat pumps and EVs. There should be a mandate placed on electricity companies to install an increasing number of heat pumps in existing buildings, as well as immediately banning the use of fossil fuels in new buildings and initiating much bigger insulation programmes.

After the changes in the system, putting a greater reliance on government intervention and public ownership, corporate interests will still be evident in this system. However, this influence can be mitigated to a great extent by consumers and activists taking power into their own hands to change the system in various ways. Incentives and funds need to be made available to local councils and community groups to innovate in energy, whether it be through developing heat networks served by large-scale heat pumps or peer-to-peer trading by community renewables companies. This activism will be important in the battle for an energy transition to replace fossil fuels and address climate change. This dimension is further explored in Chapter 5.

3

The USA and Canada:
How Pro-corporate Policies Have
Slowed the Energy Revolution

THE USA: LAGGING BEHIND OTHER COUNTRIES

The USA may be the leader of the Western world, yet it is not leading the transformation to a green economy. The proportion of energy being generated from wind and solar power lags behind many other Western economies. Whereas solar and wind power generated 22 per cent of electricity in European countries in 2022,[1] in the USA it was rather less than this, at 14 per cent.[2] In this comparison, we must remember that the list of European countries includes several relatively poor ones. Germany is relatively more affluent, and although it is not an especially windy or sunny country, in 2022 it produced 37 per cent of its electricity from wind and solar PV.[3] In Spain, 32 per cent of electricity production came from wind and solar PV.[4] In Denmark almost 60 per cent of electricity came from these two technologies.[5] Meanwhile, China's proportion of electricity from wind and solar is at just under 14 per cent and rapidly increasing.[6]

The USA has greenhouse gas emissions which, per capita, are among the highest in the world, second only to Australia in OECD countries.[7] European countries emit only around half as much carbon per capita compared to the USA.[8] So how has this situation emerged? How is it that such a rich country as the USA has not done better?

I discuss why the USA has lagged behind Europe in developing renewables. I also examine how this could change. One pressure for change could be the momentum in favour of green energy created by Biden's IRA. The IRA means that the ability to own and finance renewable energy projects is being given to many other people besides the big corporations. Second is the continued decline in the costs of solar PV and wind power. A third factor is that renewable energy's biggest competitor, natural gas, may become more expensive.

The American oil industry dominated the early years of world oil. That was until the 1970s, when a combination of tightening supply compared to fast rising demand strengthened the hand of the developing countries involved in the Organization of Petroleum Exporting Countries. Largely at the same time, a wave of nationalisations of oil industry assets reduced the power of American oil companies. Indeed, after the 1970s, with oil imports increasing and domestic US production flagging, the American oil behemoth was in decline.

Then, in the early years of the twenty-first century, a new technology – shale oil and gas (fracking) – emerged and quickly boomed. High oil prices in 2008 sparked a major growth in shale oil and gas. US oil was once more a leader in global oil production alongside, and most recently even out-producing, Saudi Arabia and Russia. These developments have probably slowed the possibilities for a sustainable energy transition in the USA.

The continued growth in the extraction of shale gas in the USA can be seen in Figure 3.1 and US oil production since World War II in Figure 3.2. Shale gas now makes up around three-quarters of US natural gas production, and its growth has much more than offset a decline in conventional natural gas extraction. One key difference between the USA and other countries is that in the USA mineral rights are the property of the landowner whereas in the rest of the world the state owns mineral rights. This has given local landowners much more of a financial interest in shale oil and gas extraction than in other countries. This might help explain how it is that fracking has taken off in the USA but not other parts of the world.

Oil and gas lobbyists have remained strong. The new, plentiful and cheap supplies of shale gas have provided new opportunities to build gas-fired power plants. Gas power rather than renewables has been replacing coal-produced electricity. A strong gas lobby

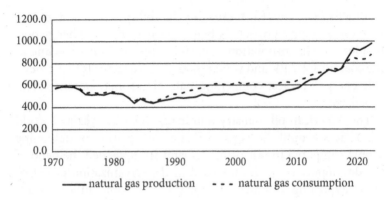

Figure 3.1 US natural gas consumption and production since 1970 in billion cubic metres per year

Source: EISRW 2023.

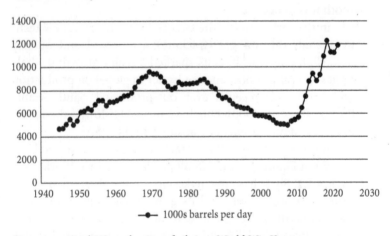

Figure 3.2 Total US production of oil since World War II

Source: US Energy Information Administration, 'Petroleum and Other Liquids – US Field Production of Crude Oil', 31 July 2023, www.eia.gov/dnav/pet/hist/LeafHandler. ashx?n=pet&s=mcrfpus2&f=a.

has also been delaying efforts to phase out the use of gas in buildings energy in favour of heat pumps. President Biden's IRA, passed in 2022, includes the financial boost for industries producing materials and parts for technologies, including electric cars and batteries. This will encourage reliance on domestic mineral sources and encourage battery recycling. It makes incentives available for renewables on a much more consistent basis than before. It makes it possible for publicly owned electricity authorities, community renewable projects and homeowners to make use of investment tax credits, which were previously only available to big companies with tax liabilities.

However, the IRA has relatively fewer funds available to reduce energy demand. Another key limiting factor is that the federal incentives do not guarantee that renewable energy developers can be given the guaranteed prices for their generation output that they need to develop the projects. This is in contrast to western Europe where PPAs with guaranteed prices for renewable output are issued by central governments to renewable energy developers. In effect, the existing energy corporations have a chokehold over issuing PPAs to renewable energy generators. The corporations can protect their own fossil fuel power plant.

Attempts by the federal government to try to mandate targets for renewable energy development across the USA have not succeeded so far. The Obama and Biden administrations have not had the congressional votes to achieve this. The Supreme Court has rejected presidential proposals that would allow the Environmental Protection Administration to compel states to meet targets for decarbonisation. The lack of ability at a federal level to influence these state-based factors may ensure a continuing bias in the system towards utilities and a pace of renewal of the grid infrastructure that is far too slow. The movement for a Green New Deal had some reflection in measures achieved in the IRA. Unfortunately, many of the aims and the scope of the Green New Deal were frustrated because of push-back pressure from pro-fossil fuel interests.

There are several areas that can be looked at to investigate the issue of the relative underperformance of green energy in the USA. This includes the nature of the energy regulatory system at state and federal level, and how the system favours the big energy corporations and fossil fuel and nuclear interests to the detriment of independent renewable energy developers, energy efficiency and rooftop solar PV.

Given that much of the action is state based, it is important to look at particular states. There is a special focus on one state,

Florida, to illustrate some of the issues involved. Finally, I discuss energy outcomes in Canada.

REDUCING ENERGY DEMAND THROUGH EFFICIENCY

Around four-fifths of energy came from fossil fuels in the USA in 2021. Energy efficiency can provide a large proportion of the work in substituting fossil fuels. Electrification is the essential pathway to reducing energy demand, and this includes many industrial energy uses, as well as cooking and heating in the home. Transport accounted for 28 per cent of US energy use in 2021,[9] and the bulk of this is used in road transport.

Since the 1970s there have been legislative efforts to improve the energy efficiency of motor vehicles, using the Corporate Average Fuel Economy regulations. However, these have been weak in effect, and flouted by the growth of sports utility vehicles (SUVs), which are not covered by the regulations. The switch to EVs will automatically improve fuel energy efficiency, although sadly many of the new EVs are SUVs. SUVs are regarded as a blight by environmentalists. Sales of EVs in the USA have been relatively lower than in the EU or China.

Campaigns for greater emphasis on walking and cycling are often more challenging compared to Europe. US cities were designed for the motor vehicle, much more than denser, older, European cities. I recall how, on a visit to Baltimore, I wanted to go to a restaurant no more than two miles from the hotel. 'I'll walk', I told the receptionist, who looked at me in horror. 'Walk? You can't do that. I'll call you a taxi', they said. I waived the offer aside and set off. Even though the area was heavily urbanised, I discovered that there were usually no sidewalks (or pavements as us Brits call them). I must say it made for difficult going. I got a taxi on the way back.

As discussed in Chapter 1, heat pumps can reduce the energy used for heating and cooling by around 70 per cent. New buildings need to be constructed with far tougher insulation specifications to reduce heating needs in winter. Buildings also need to incorporate ventilation systems that reduce the need for cooling. Well-designed buildings need very little energy for heating. Air conditioning systems that use less energy for a given amount of cooling need to be more vigorously promoted. In 2021 the Rocky Mountain Institute in collaboration with the government of India and Mission Innovation announced the winners of a prize for designing super-efficient air conditioners whose operation reduced the climate impact of cooling buildings by fivefold.[10]

New buildings need to have building codes that improve energy efficiency dramatically. There has been an uptick in US interest in low-energy building designs such as passivhaus in order to emulate some progressive cities in Europe.[11] The USA was also well behind the EU in improving lighting efficiency. The EU introduced laws effectively banning incandescent lightbulbs in 2009, but this was not done in the USA until 2022.

Besides the need to switch to heat pumps for heating and cooling, we need to switch from gas cooking to electric cooking. In this way, indoor air pollution, from burning natural gas, is also avoided. The Consumer Product Safety Commission has been considering a ban on new gas stoves, in order to protect consumers' health. Environmental activists have been campaigning for banning the use of new gas appliances and also banning gas connections to new buildings, with different results in different parts of the USA. In states such as New York and Washington, as well as in various cities across the USA, plans are advanced to ban natural gas connections in new buildings and heat pump installation is being incentivised.

I do not believe it is really the case that Republicans have some ideological connection to continuing to use fossil fuels, but it is the case that conservative, Republican politicians are more vulnerable to appeals from fossil fuel interests. The Sierra Club, the leading environmental group in the USA, has attacked the natural gas industry for being behind attempts to 'ban the bans' on new gas connections.[12,13] In Florida, for example, a law was passed in 2023 under which local governments are being stopped from banning gas stove installations in new buildings. The aim of banning gas stoves is to protect people from nitrogen dioxide emissions that can cause lung problems. It also reduces carbon emissions since the alternative is electrically powered stoves.

Improving energy efficiency is the most important thing in reducing reliance on fossil fuels, yet there is a lack of federal leverage on this. There is little at a federal level to promote heat pumps or super-energy efficient buildings. A federal electrification programme is badly needed. Since a lot of energy efficiency is about electrification, there is a natural marriage with renewable energy which is mostly supplied as an electric source. More electrification dramatically expands the market for renewable energy. There has been a notable uptick of sales of air source heat pumps in the USA in recent years, as can be seen in Figure 3.3. According to New York State, 2022 was the year that sales of air source heat pump exceeded sales of gas furnaces (which are used more widely than gas boilers in the USA).[14]

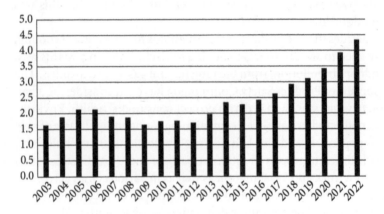

Figure 3.3   US air source heat pump sales (millions) since 2003
Source: The Air-conditioning, Heating and Refrigeration Institute (2023)
'Air Source Heat Pumps', www.ahrinet.org/analytics/statistics/historical-data/central-air-conditioners-and-air-source-heat-pumps.

WHAT POWERS ENERGY IN THE USA?

In order to understand US performance on green energy, it is helpful to briefly outline some of the most relevant institutions. As mentioned earlier, the USA has one of the highest per capita rates of energy use and carbon emissions in the world. Three factors are influential here. First is the sheer size of the nation's economic output relative to others (per capita). Second is the dispersed nature of US settlement and the relatively expansive space taken up by buildings. As a consequence, a lot of energy is used in transport and also energy services for buildings. Third is the nature of the policies affecting energy which, on the one hand, do not do enough to improve the efficiency of energy use, and on the other hand do not do enough to encourage renewable energy.

Since building codes and other regulations affecting buildings are state based, this means that regulations influencing energy efficiency are also largely state based. As discussed already, energy efficiency is such an important part of the energy revolution that it is a major problem if energy efficiency is left up to piecemeal progress at the state and local level. This underscores the importance of well organised, locally based activism. Some states leave building codes entirely up to cities. This makes campaigns for energy efficiency more difficult to organise and focus.

The answer to why the USA has been falling behind on renewable energy may also be influenced by the types of companies that dominate the US electricity market. Among the companies that run electricity operations in the USA, the majority are called 'investor-owned utilities' (IOUs) (i.e. private, for-profit operations), while the others are municipal companies or cooperatives. Zach Stein says that: 'According to the U.S. Energy Information Administration (EIA), IOUs served 72 percent of U.S. electricity customers in 2017.'[15] The publicly owned municipal electricity companies have, up until the passage of the IRA, been unable to invest directly in renewable energy projects, and very often the IOUs have protected fossil fuel energy production.

The main policies, promoting renewable energy, have come in two parts. First, at a federal level, renewable energy (and nuclear power) is, and has been, promoted through a system of tax incentives which reduce the costs of developing renewable energy schemes. Second, at a state level there are renewable portfolio standards (RPSs) which produce mandates to secure a stated amount of electricity from renewables by a given time. These two policy instruments should, in theory, interact. However, RPSs are often either too weak to promote continued renewable energy development or non-existent.

WHERE US STATES HAVE DONE WELL IN RENEWABLES

California is well known for its 'windrush' in the 1980s, something that came off the back of a growth in alternative environmental and energy movements in the state in the 1970s. The California Energy Commission (CEC) was established to design strategies for sustainable energy.

In Jerry Brown's first incarnation as governor, a raft of supportive tax measures were set up to support renewable energy development, and the utilities were prevailed upon to issue long-term contracts to pay developers that were broadly similar to 'feed-in tariffs'. Although the programme was ended after a few years, the push for renewables in California had begun. In 2021, California derived around 38 per cent of its electricity from wind, sun or geothermal energy sources. Recently the most growth has been in solar PV, although the CEC is planning for a major programme of offshore wind farms. However, as in other parts of the USA, developers face having to make multiple applications for permits for offshore wind farms to different agencies. This contrasts with the 'one-stop' planning regimes that deal with offshore wind in countries such as Germany and the UK.

California leads the USA in terms of the proportion of its electricity derived from solar PV, and is way ahead in that respect compared to other states with the best solar resources, such as Nevada, Arizona and Texas.[16] However, California has relatively few areas of high wind speeds and a relatively high population density. This leads to fewer opportunities for wind farm development compared to states with lower population densities and greater quantities of wind resources, such as in the Midwest.

Wind power has expanded quickly in Iowa in recent years. Despite having better wind resources than average for the US states, Iowa does not have the wind resources of other Midwest states such as South Dakota, North Dakota, Kansas and Nebraska.[17] Nevertheless, Iowa leads the USA in the proportion of electricity derived from wind energy by any state. This amounted to 58 per cent in 2021. It was 'populist' pressure from below that forced the change which kicked off Iowa's wind power surge. According to a pro-renewable energy activist, Ed Woolsey: 'The populism came from the ownership aspect, smaller farms and small co-ops, actually being able to provide for their own energy and actually the ownership option there of being able to afford to enter into this new energy industry.'[18] The monopoly utilities had to be squared, however. This was done when Governor Tom Vilsack made a deal in 2000 with the electricity utilities that they would make extra profits on the electricity infrastructure associated with wind power, including transmission lines.[19] Resources do count of course, which is why windy Wyoming, despite having strong pro-coal lobbies, has achieved high penetrations of wind power, with 22 per cent of its electricity coming from wind power in 2021.[20]

Where renewable energy has advanced most rapidly, both within and outside the USA, there are common themes. First, there have been popular campaigns involving farmers, cooperatives and environmental NGOs. Second, there has been strong intervention at the state political level to ensure a consistent regime supporting renewable energy. This tends to be at least as important as the relative quantity of different renewable energy resources that may be present within a state's jurisdiction, otherwise big energy corporations/utilities will simply carry on with their old technologies and policies. They will adapt only slowly, if at all, to changing public demands for greener energy provision. That being said, where electricity demand is increasing, the utilities are nowadays much more likely to choose renewables because they will deliver energy more cheaply than gas-fired or coal-fired power plants. The combination of activist pressure for support for renewable energy, and state action to ensure that dominant utilities allow renewable energy development, has been sadly lacking in the majority of US states to a greater or lesser extent.

LOW INCENTIVES THAT FAVOUR CORPORATE INTERESTS

The US federal government has been giving incentives to renewable energy since 1992 in the form of the 'Production Tax Credit' (PTC) or 'Investment Tax Credit' (ITC), depending on whether the tax credit is given in regard to the production of energy or investment in renewable energy equipment. Under the IRA, the PTC amounts to 2.6 cents per kWh in 2022 prices – or a 30 per cent reduction in costs with the ITC (with 10 per cent extra for meeting local content criteria).[21] The PTC and ITC support has been dependent on congressional renewal over the years. The mode of giving this incentive through tax incentives is different to most other countries. The most popular means, certainly in Europe, has been through governments offering fixed price contracts. This entails the developers being awarded guaranteed prices for energy generated over a fixed period, usually 15–20 years.

In the USA, the level of the PTC incentive for renewable energy, 2.6 cents/kWh, does not help developing technologies. More expensive technologies do not usually obtain extra incentives. In order to help offshore wind and solar PV develop (and then lower costs), national European governments allowed high prices to be paid to developers of these technologies. Criticisms were made of these high levels of incentives, but they proved to be very effective, in that bigger markets were created. This led to rapidly declining technology costs. In the USA, it is up to the individual states to add extra incentives for new technologies. Yet this piecemeal approach has been ineffective. Essentially, the USA has relied on other countries to do the heavy lifting in terms of optimising renewable energy technologies.

No doubt the US system has had some key disadvantages that have reinforced corporate power. Paul Gipe, a longstanding campaigner for community-based renewables, said of the PTC system that 'only the most profitable companies in the nation – those with the most taxes to pay – could participate in the program. Homeowners, farmers and small businesses need not apply.'[22]

The PTC system has also meant that publicly owned municipal utilities and rural electricity cooperatives have usually been unable to invest in renewable energy since they do not have a tax base which they can offset to use the PTC. The only way to access renewable energy is by making PPAs with privately owned companies. This makes the renewable energy more expensive. However, the IRA has altered the incentive conditions, so that now entities that do not have a tax liability can obtain the PTC benefit through a direct cash payment. This should make a big difference to publicly owned electricity entities.

Because of the need for extensive tax liabilities, the PTC system, according to analyst Sarah Knuth, 'mostly benefits big banks'.[23] After conducting wide-ranging research on the subject for a top environmental journal,[24] she concludes:

The pool of tax equity investors is scarce compared to the number of renewable power projects seeking their capital. Frequently, banks make developers pay them sizeable fees for their participation. They also command outsize power in determining which projects get developed, and by which developers. Because banks profit more from big deals, they prefer the largest private developers and mega-projects. The average renewable tax equity deal is US$150 million, and offshore wind developers may soon require as much as US$800 million per project. Meanwhile, smaller competitors and projects often get no deal at all.[25]

The low level of incentives favouring renewable energy is to be compared with the large sums offered to fossil fuel interests, for CCS projects, in the IRA. This technology, as mentioned in

---

**Box 3.1   Fossil fuel carbon capture:
a failing North American technology**

As a column I wrote for the website 100percentrenewableuk.org puts it:

CCS technology is a failing technology. The only power stations that have used this technology have demonstrated CCS hopelessness. At best the technology is hoped to capture almost 90 per cent of carbon dioxide, but in reality, the only two major demonstration schemes involving power plants have never achieved anything like that. The longest-running project, the Boundary Dam coal-fired power plant in Canada, worked so badly that less than half of the carbon dioxide has been captured.

The other major demonstration scheme at a power station to have been tried is, Petra Nova, in Texas. This has achieved a carbon capture rate of no more than 65–70 per cent.

Quite apart from the failure of the technology, three unanswered questions hang over the use of CCS in power plants. The first is whether a technology that, even if it worked properly, could not deliver more than 90 per cent carbon capture, was something that should be a target for Government subsidies. Wouldn't spending on innovative renewable energy technologies such as tidal stream and wave power, not to mention more established renewables and energy efficiency, be a much better bet? Why not put some money into developing fuels made from renewable energy that could be stored? That would do the job that CCS is supposed to do anyway.[26]

Chapter 1, is not totally effective in removing carbon from power plants anyway, even if it operates as well as planned, which so far it has not. It would be preferable if the money was given to renewable energy and energy efficiency schemes. However, the politics of the pressure from fossil fuel interests, and the need to get the IRA passed, meant that funding for CCS is very generous even though CCS is not a sustainable alternative to renewable energy. Box 3.1 talks about fossil fuel capture as a failing technology.

LOW INCENTIVES THAT DO NOT ENCOURAGE INNOVATION

There has been little opportunity for the development of new technologies that require more incentives at the start of their development. In the USA, the so-called free market ideology, which prescribes against giving high fixed prices for renewable energy generation, in fact just preserves existing dominant technologies. The appeal to free market ideology (which rails against price-setting) is a convenient argument that suits those invested in fossil fuel and their friends in energy corporations. It is also one that seduces right-wing politicians to support their preferences.

The USA has, for the most part, relied on the development of markets in Europe that have allowed technologies to achieve technical optimisation and economies of scale. The US approach to offshore wind has so far been feeble in comparison with Europe. Offshore wind is now very cheap in the North Sea, and part of this is down to the fact that major developments by several countries, especially the UK, have generated economies of scale in the industry.

The US approach to offshore wind, for example, has failed to consolidate planning rules, leaving developers to seek a number of different permits. Perhaps even worse, the funding has been inconsistent, depending on different states. Regulators have intervened with shifting sets of rules and demands, and contracts offered by utilities have been variable. The absence of a federal system of awarding premium price contracts for offshore wind meant that, for instance, opponents of the Cape Wind project in Massachusetts were able to challenge the funding offered to the project, thus delaying it. Then the utilities, which had offered PPAs for the project, decided not to extend them to take account of the delays, ending its viability.[27] Fishing interests have opposed offshore wind developments, but they have been supplemented by opposition funded by fossil fuel interests.

REGULATORY BIAS AGAINST RENEWABLES

A crucial part of the US energy system is the regulatory framework. There is a Federal Energy Regulation Council that deals with interstate and matters covered by the Public Utility Regulatory Policies (PURPA) laws. At the state level, the energy systems are regulated by 'Public Utility Commissions' (sometimes called Public Service Commissions or State Commissions), which have important powers in authorising investments, profit rates and consumer prices. The nature of their responsibilities, for the electricity sector in each state, varies according to the degree of market liberalisation (allowing competition). However, a common theme is that the state regulatory agencies have sided with the utilities, that is big energy corporations.

There have been criticisms of 'revolving doors' between the power companies and the regulators.[28] These links between regulators and corporate energy companies can raise questions about how decisions made by regulators will favour the technological status quo over new technologies such as batteries, solar and wind farms, residential solar PV or DSM techniques. According to S&P Global Market Intelligence, out of 54 regulatory commissions, 'members are appointed in 39, while members are elected in some fashion in 15'.[29]

Indeed, sometimes state officials have been more conservative than the state's own energy corporations. In the case of West Virginia, plans for wind farms put forward by its leading power company, Appalachian Power, were blocked by the regulators. According to a story published in 2018, the 'State Corporation Commission in April rejected a request by Appalachian Power to buy 225 megawatts of wind capacity, saying that the utility failed to establish a need for new generation to serve its Virginia customers'.[30]

Even in states with a relatively good record in bringing wind power on line, such as Iowa, there is evidence of regulatory bias against renewables, promoted by independents, and in favour of incumbent corporations. The Iowa regulators ruled against a plan for an independent company to provide solar power to the City of Publique, although this ruling was later reversed by the District Court.[31] The regulators have rejected efforts by local people to create local municipal electricity utilities that could invest more heavily in energy efficiency and renewable energy than the incumbent energy corporations were prepared to do.[32]

There has been commentary about how state regulators have generally used their interpretations of economic criteria to hamper efforts to implement state mandates in favour of rapid moves towards renewable energy. According to an article in the *Harvard Law Review*:

> PSCs (Public Service Commissions) are creatures of habit and have developed case law, administrative procedures and staffing decisions for a century through an economic lens. This narrow focus is due to the PSC's traditional economic mandate to hold in check the monopolistic market power of utility companies and serve as a proxy for real world competition. ... While a few states have successfully embraced their role in the climate solution, they stand as the outliers. Most state PSCs remain entrenched in their traditional economic mandate, refusing to consider the impacts of their energy decisions on the climate and, at times, undermining the will of their electorate.[33]

Regulators may claim to adopt an apolitical approach, favouring consumer interests in keeping prices low. However, in doing so they are pursuing an implicit ideological bias, putting what they call 'competition' ahead of climate objectives. Really what they are doing is not favouring competition but the technological status quo. Ironically, when they obstruct proposals for renewable energy, they are in fact confirming monopoly corporate interests, which have been threatened by climate mandates demanded by the electorate.

The dominant US regulatory approach protects existing conventional power station assets from competition from new renewable projects. Existing projects have already seen their construction costs paid off. This means that new projects can only, under this policy (e.g. as in the West Virginia example given earlier), be given the go-ahead if they can be seen to cost less than the operating costs for the conventional plants. This policy will restrict renewable energy to those states where there is a rapid increase in electricity demand, requiring new power plants to be built. The protection of existing conventional power plants will tend to protect the interests of the dominant energy corporations against competition from independent renewable energy developers. Leah Stokes comments that, 'throughout the South uneconomic coal plants continue to operate despite the availability of cheaper and cleaner alternatives. Since utilities have sunk debt and equity into these plants, they want to keep them open.'[34] However, even in the more northern state of Ohio, coal interests have weakened regulations supporting renewable energy growth.[35]

Leah Stokes has detailed how and why fossil fuel lobbyists successfully limited the level of RPS support for renewable energy in Texas. Renewables were seen as a threat to the interests of natural gas generators, because renewable energy schemes would reduce the prices at which the power from the natural gas power plant could be sold.[36]

In Texas, the electricity blackouts in early 2021 were falsely weaponised by fossil fuel supporters against wind power. According to an academic analysis, the cause was lack of weatherisation of power plants in general: 'Texas failed to sufficiently winterize its electricity and gas systems after 2011. Feedback between failures in the two systems made the situation worse. Overall, the state faced outages of 30 GW of electricity as demand reached unprecedented highs.'[37]

The probability of blackouts is made more likely by poor grid infrastructure. Added to that, many renewable energy projects are stalled because of the need to upgrade the grid to accommodate them. Among the factors listed as slowing efforts to upgrade the grid is that, according to a special Reuters report, 'utility companies often fight investments in transmission network improvements because they can result in new connections to other regional grids that could allow rival companies to compete on their turf'.[38] Although there is an impressive number of solar and wind farms waiting in the queue for grid connections, actual projects are likely to be much fewer.

Even in supposedly green energy-friendly California, the regulatory authorities are supressing independent initiatives to install green energy that conflict with corporate utility interests. The California Public Utility Commission is supporting efforts by monopoly electricity providers to stop independent companies from establishing microgrids. Microgrids can make local communities self-sufficient using renewable energy, and as such may be an important part of the green energy revolution.[39]

ATTEMPTS TO GIVE PRIORITY TO NUCLEAR POWER

Nuclear power supplies 20 per cent of electricity in the USA. This is based on nuclear power stations that were built from the 1960s. That was at a time when electricity utilities, with encouragement from the federal government, were building nuclear power stations as what was seen as the energy supply of the future. However, the incorporation of basic safety measures, such as containment walls, increased nuclear costs. Designing larger power plants was intended to offset such problems, but the economics did not improve. Following the 1973 oil crisis, utilities were under pressure to reduce costs, as electricity demand was not increasing as fast as had earlier been anticipated. Since building new nuclear power plants were seen as a cost burden rather than a cost saver, future plans for nuclear plants were cancelled.

There are now just under 100 reactors in the US nuclear fleet, and they are being retired, as the cost of refurbishing them outstrips their ability to generate income. However, nuclear operators have managed to increase the performance of the old reactors to produce electricity. The outcome is that the total electricity production from US nuclear plants has remained roughly constant this century.

Regulators have allowed utilities in some states to push incredibly large resources in the direction of developing new nuclear power, under terms of which renewable energy developers could only dream. Nevertheless, the results have been meagre. In the case of nuclear power construction, cost overruns have been automatically paid by ratepayers (energy consumers) and US electricity utilities have a natural liking for nuclear power as opposed, in particular, to solar PV. As one campaigner for renewable energy said:

> So you have an enormous influence from the big companies that have a captive audience and their business models, so get a guaranteed range of rate of return on their capital expenditures. Because the incentives for these utilities is the bills ... then they get a guaranteed rate of return ... if you're a big monopoly utility and you've been operating on this business model, you want to build a big centralised power plant, but (independently owned) solar really threatened that.[40]

One of the problems with renewables' policy over the last 20 years has been the false hopes that nuclear power would provide the shift towards non-fossil energy that was required. Hence policy attention has been diverted away from promoting renewable energy more vigorously. Indeed, in 2007, one of the top officials of the Nuclear Regulatory Commission, Jeffrey Merrifield, predicted that 'in the next twenty years, assuming continued safe operation, we could at least double the number of nuclear power plants we have in this country'.[41] In fact, despite the incentives being available for new nuclear plants, the number of nuclear power plants has fallen in the USA since 2007, and only two plants which began construction this century are being completed.

In practice, only Georgia seems to be bringing any nuclear plants to completion, albeit at great cost and delay. In Florida, South Carolina and Georgia the state regulatory authorities allowed the (utility) developers to recoup their construction cost expenses before any generation (and therefore income from generation) occurred. Florida Power and Light (FPL) planned to build two reactors at Turkey Point and began cost recovery for its expenses in doing so, taking nearly $300 million from ratepayers. The project has been indefinitely postponed since 2016. In South Carolina, construction of two large nuclear reactors at VC Summer began in 2013. Cost overruns and delays overwhelmed the project. West-inghouse, the manufacturer producing the reactors, went bust in 2017, leaving South Carolina ratepayers on the hook for several billion dollars. The debacle was so bad that several officials of the SCANA Corporation energy utility (which went bust itself) were charged with criminal offences. In 2021 the ex-CEO of SCANA was sentenced to two years in jail for fraud relating to the scandal.[42]

A similar project for two large nuclear reactors at Vogtle in Georgia also started construction in 2013, and the first reactor finally came online in 2023. This project has been a long-run-ning saga of delays and cost overruns. The reactors were also being supplied by Westinghouse, but the project survived the company's bankruptcy in 2017. The main owner, Georgia Power, decided to continue the project, aided by what has amounted to guaranteed federal loans of $12 billion. The project is estimated to be costing its owners more than $30 billion in 2022 prices. This is over twice as much as originally projected at the time the project was com-missioned.[43] Again, this project has been financed out of a cost recovery mechanism for which ratepayers pay through their elec-tricity bills, long before the plant actually supplies any electricity. According to Georgia Conservation Voters, Georgia Power is making $6 billion out of the development.[44]

There are no plans to build any further large reactors in the USA. Rather, there has been a growth in hype about the much fan-tasied (and frequently revisited in different forms) 'renaissance' of nuclear power now taking the form of SMRs. As discussed in Chapter 1, there seems little hope of a substantial industry arising from this so-called group of new technologies. It seems unlikely that the supposed cost benefits of making small nuclear reactors will offset the disadvantages of losing economies of scale.[45]

REPRESSING SOLAR

Large-scale solar power is deterred by the US electricity system because the utilities make profits out of its power plants, not solar (or wind) farms put up by independent companies. However, rooftop solar has been especially well repressed to the advantage of the utilities and fossil fuel interests in general. This occurs in several ways. First, the bureaucracy facing would-be solar PV home installers is Byzantine in its complexity by comparison with European or Australian standards. Second, the rules about investing in installing solar PV themselves favour the utilities. Third, in many states at least, solar panel owners are not paid decent rates for the electricity sent to the electricity system.

I have been astonished how expensive solar power on rooftops is in the USA compared to Europe. For example, in March 2022, for £8,250 I had 6.2 kW of solar panels and also a 5.1 kWh battery installed in the house into which we had recently moved. That is around $10,000, or, once you have taken the price of the battery off, about $1.20 per watt in the money of the day. But in the USA, prices (in 2022) were much higher – at about $2.94 per watt.[46] Of course, under the IRA the tax credit now brings down the cost that the consumer pays somewhat, to about $2.20 per watt. So the full price of a US solar installation is over two and a half times more expensive than what I paid in the UK, and even with the inclusion of the tax credit it is still nearly twice as expensive for the consumer compared to the UK. (In the UK there was also a feed-in tariff system to support solar PV, but this stopped in 2019.)

An article by Todd Woody in *Forbes Magazine*, in 2012, talked about how solar installation prices were twice as high in the USA compared to Germany, and it seems that little has changed since then. Woody comments:

Studies by the National Renewable Energy Laboratory and by the University of California, Berkeley both confirm that these higher prices are almost exclusively related to the paperwork it takes to 'officially' install a standard rooftop system in the U.S. That's right, government red tape – local, state and federal. ... In Germany, the residential solar industry has no red tape, there is a highly-tuned supply chain to get equipment to customer job sites, installers get projects completed in a day, permitting is virtually automatic, costs to acquire a customer are very low and overhead is negligible. ... The resulting complexity in the U.S. means that the solar salesperson needs to be a highly trained spreadsheet expert and it takes specialized engineering to design systems to be compliant with the plethora of local building/fire/

utility codes and even the smallest solar installer needs a full-time employee just to process all of these documents.[47]

All of this suggests that the tax credit for residential US solar is not even paying for all of the excess bureaucracy involved in the US solar installation system.

Solar PV installers have to deal with unhelpful technical requirements for connections to the electricity system, and utilities certainly have no incentive to influence any change to this. The utilities' interests are opposed to consumers' self-generating power. The more this happens the less revenue is earned by the utilities. The utilities have very large financial resources to give money to politicians, which may be paid for out of the rates paid by the consumers. The consumers with solar rooftops cannot make large donations to politicians. They do have advocacy organisations such as Vote Solar and the Southern Alliance for Clean Energy to argue for them, but their resources are thinly spread and meagre compared to those of the utilities. As an article in *Grist* put it:

> Working with trade groups like the Edison Electric Institute and the American Gas Association, utilities have blocked policies that would promote rooftop solar and electrify buildings. At the same time, they've kept the country hooked on fossil fuels by, for example, paying Instagram influencers to defend gas stoves. Nearly half of the 25 largest investor-owned utilities in the United States are working to delay climate action, a report from the think tank InfluenceMap found last year.[48]

Roughly half of US states have had net metering that allows consumers to recoup generation that they send to the grid at the retail electricity rate.[49] 'Net metering' means that owners of solar panels are paid for power exported to the grid at the same rate that they would pay for the electricity they consume. Some states still have restrictive laws on whether 'third parties' can finance rooftop solar installation, and some have a legal ban on this practice altogether.[50] Organisations such as Vote Solar and the Solar Energy Industries Alliance have fought many pitched battles with utilities who have favoured restricting the conditions under which solar PV is installed. The issues have involved third-party financing, trying to fend off efforts to tax solar installations and arguing about the amount that solar generations will be paid for the energy they send to the grid.

In California there is an argument over the extent to which rooftop solar or large solar farms should be encouraged, with conservationists opposing the development of as yet untamed desert resources. Do all desert resources have to be ruled out? Certainly, the bulk of solar PV installed so far is utility scale rather than rooftop based. Yet regardless of these arguments there are possibilities for solar farms on land which has already been converted to farmland and for which water resources for irrigation are becoming depleted.[51] In addition, there are legislative efforts to harness transportation routes to enable solar deployment.[52]

Organisations such as the Southern Alliance for Clean Energy argue that as the amount of rooftop solar increases in a particular electricity network, there should be targeted 'net metering' incentives for rooftop solar PV. This should be aimed at encouraging solar generators to use home batteries and flexibility techniques. Batteries and flexibility techniques, such as the use of thermostats, can postpone electricity consumption to periods outside of peak consumption times. This could help stabilise the grid and thus earn value for all consumers. It could also get around claims that net metering is a subsidy for rich consumers at the expense of others. What is certain is that eliminating the unnecessary red tape that surrounds the installation of solar PV would reduce the need for the system to give incentives to solar PV generators.

Figure 3.4 shows the relative number of residential solar installations compared to an average 100 persons living in selected different Western countries. The USA has the lowest record. Germany and the Netherlands have had the most consistent policies favouring solar PV on rooftops and a simple set of regulations governing their installation. The USA has neither. The Netherlands has had a consistent set of full net metering since 2004 which entails home solar generators being paid the full retail amount for any electricity sent to the grid, and Germany has had a consistent feed-in tariff in place for residential solar PV. The UK has had much less consistent financial support for solar PV, although there are fewer regulatory restrictions on solar PV compared to the USA.

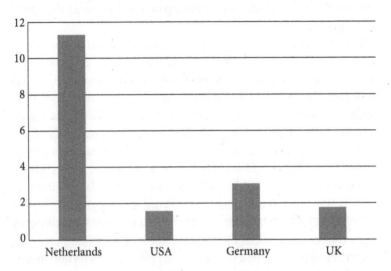

Figure 3.4   Solar roofs per 100 people in the USA and other selected developed countries

Sources: Worldometers population data; Clean Energy Wire; Statista, the ecoexperts; Powermag.

In the USA, utilities have often promoted legislation to restrict solar PV. One of the most notorious episodes was an attempt by FPL (in 2016) to organise a ballot initiative in Florida named 'consumers for smart solar', which in fact denied the right for third-party financing. In the end the initiative failed to gain sufficient votes. After a protracted struggle, solar campaigners managed to gain a reduction in taxes levied on solar panels. Ways around the ban on third-party financing were found, and after this the installation of solar rooftops took off in Florida.

In April 2022, Governor Ron De Santis vetoed a Bill supported by FPL that would have allowed FPL to increase the rates paid for electricity consumers to compensate for its loss of revenue caused by solar rooftop installations.[53] This was a political turnaround, since FPL had been accused of funding a candidate to help the Republicans retain the state Senate in 2018 elections.[54] However, in recent times, perhaps influenced by the larger company which owns it (NextErA Energy), FPL has been building a lot of solar farm capacity. A lot more solar PV is planned in Florida.

HOW FOSSIL FUELS CONTINUE TO GET THE BEST DEALS

A 2017 Congressional Budget Office Testimony (looking at incentives for domestic US production)[55] calculated that incentives for fossil fuels were, at that time, about 40 per cent of the level of the incentives then being given to renewable energy. On the other hand, the Budget Office Testimony says: 'From the introduction of tax preferences for oil producers in the Revenue Act of 1916 until 2005, the largest share of energy-related tax preferences went to domestic producers of oil and natural gas'. Hence in the past fossil fuels have been given many times more tax compared to renewables. The report 'concluded that eliminating the preferences would reduce domestic oil production by less than one-half of one percent and would have virtually no effect on the domestic price of gasoline'.[56] The Obama and Biden administrations have talked about eliminating fossil fuel subsidies, but there is as yet no sign this is happening.

Meanwhile, natural gas power plants are still being built in the USA in large numbers. A total of 27 GW was reported as coming online in the 2022–5 period. The bulk of these are in the states of 'Illinois, Michigan, Ohio, and Pennsylvania' and also Florida and Texas.[57] If incentives for renewable energy were good enough, this would not be happening. There is perhaps even lesser awareness of the fact that, as time goes on, the domestic US shale gas industry may not keep natural gas prices down.

The USA produces a lot of oil, but this does not help American consumers of oil products, to the extent that oil prices and refined oil product prices are globalised. The same could happen to the prices paid by US natural gas consumers, which could rise to much higher global market levels. This is because shale gas producers want to fill a growing gap in the global LNG market. The growth in exports of LNG from the USA can be seen in Figure 3.5. Indeed, in the summer of 2022 the Industrial Energy Consumers of America Group claimed that US LNG exports were increasing US natural gas prices. This was because the temporary closure of an LNG export terminal at Freeport, which resulted in a stoppage of LNG exports, coincided with a period of lower LNG prices.[58]

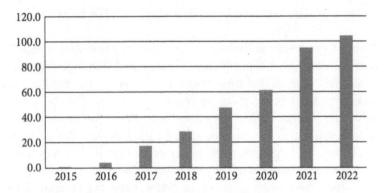

Figure 3.5   US exports of liquefied natural gas (billion cubic metres)
Source: EISRW 2023.

The US oil and gas industry claims to be a great boon to the US consumer, but the reality is that they make surprisingly little difference to the prices that consumers may pay for their energy services. Yet issues of energy security always seem to be interpreted in their favour. This has happened, for instance, in the case of restrictions being placed on foreign oil imports in the 1960s, and more recently there being no restrictions on the ability of US shale gas producers to export their products in the form of LNG.

HOW THINGS COULD TURN AROUND FOR GREEN ENERGY IN THE USA

Global pressures are likely to increase the economic pressure in favour of renewables. Natural gas prices are likely to increase and renewables costs will fall. As argued earlier, increases in US natural gas exports are likely to link the North American natural gas market to the global gas market where prices are volatile. So far the global LNG market has only made up a relatively small proportion of the total natural gas market, the bulk being transported by pipelines. However, this is changing, and has been given a big push by the pressure to replace Russian gas supplies to Europe. Natural gas markets could come to more resemble global oil markets where the majority of oil is shipped rather than sent through pipelines. Just as US prices reflect global oil prices, so North American gas prices could increasingly trend towards global LNG prices.

The second factor is the continuing decline in the costs of renewable energy. As discussed in Chapter 1, a doubling of production output of a particular technology might be associated with a 20 per cent reduction in cost of that technology. The volume of renewable energy capacity is likely to expand several times over in the coming years. This means that solar and wind power costs are going to continue to decline steeply. Renewable energy technology will outcompete fossil fuels which themselves will be increasing in price because of increasing natural gas prices. Heat pump installations will also benefit from an increase in natural gas, their main competitors.

CANADA

Canada is the largest consumer of energy per person of the major Western states.[59] This is even more so than the USA. In greenhouse terms this is mitigated to some extent by the fact that over 80 per cent of its electricity comes from non-fossil sources, with hydroelectricity providing a large share of total energy. However, despite this, Canada is not far behind Australia and the United States in per capita greenhouse gas emissions.[60] This is partly because, as in the case of these other two countries, its settlements are widely dispersed and its buildings are large. It is also cold in the winter in Canada.

This means a great deal of oil and natural gas is used in transport and heating. On top of this, an important factor is that there is considerable fossil fuel use, because of its industrial activities in mineral mining and processing (including oil production). Ontario is the largest province in terms of population, and together with Quebec, British Columbia and Alberta these provinces make up around 85 per cent of Canada's population.

Production of oil has boomed this century, and Canada has become the world's fourth largest oil producer, with around 6 per cent of global production. This is mostly from production from tar sand, that is sandy rock containing concentrated oil deposits called 'tar'. This is principally from Alberta. Its environmental effects have been loudly condemned by conservation groups and representatives of indigenous peoples. This is because of the large amount of toxic waste that the tar sands oil industry generates and the amount of land that is spoiled.[61] Like the USA, it could be argued that a lack of environmental action, on issues including climate change, could be influenced by the importance of oil to the economy. In 2011 the Canadian government even pulled out of the Kyoto Protocol. Increasing production of oil was associated with increasing Canadian greenhouse gas emissions resulting from increased use of energy in the industry and emissions from the oil industry itself, as well as a growth in the use of SUVs. Fears that compliance costs would be very high for Canada formed a large part of the explanation for this outcome.

However, more recently the Canadian government signed up to the Paris Agreement in 2015. The Canadian authorities, whether at a federal or usually also at a provincial level, have not (until very recently at least) pursued an aggressive decarbonisation strategy. The Trudeau government has tried to introduce a carbon price policy, but this has not reduced greenhouse gas emissions.[62] In 2023 it promised to introduce new regulations to achieve a 100 per cent low-carbon electricity system by 2035.

In some ways, Canada's energy policies are now much closer to that of the USA than Europe. Its policies for energy efficiency are largely in the hands of the provinces and territories, since they legislate the building codes. There is a national standard set, but there is a lot of catching up to do on the latest standards at the provincial level.[63]

Around 60 per cent of electricity in Canada is generated by hydroelectricity, with wind power making up 5 per cent of total generation. Around 15 per cent comes from nuclear power, generated by the distinctive CANDU (Canadian deuterium reactors). Ian Fairlie has argued that these reactors are prone to excessive tritium releases which pose a significant health risk.[64] Around 18 per cent of electricity comes from fossil fuels, especially natural gas.

Procurement of electricity generation is also done at a provincial level. The federal support mechanisms for renewable energy have been meagre in the past, with only 5 per cent of electricity generated from wind or solar. However, in the 2023 budget, the federal government introduced a US-style ITC at 30 per cent, the same level as in the USA. Certainly, the Canadian government is aware of Biden's IRA and the influence of the argument that it needs to match the US's green energy investment in case industry moves south. Canadian policies for renewable energy consist of tax incentives for renewables offered at a federal level, but on the other hand the issuing of contracts for energy generation are the prerogative of provincial governments and their agencies.

Traditionally, the country's largely hydro-based electricity system was built and owned by state-owned companies. However, since the 1990s, the share of power generated and distributed and supplied by private companies has been growing. In Ontario, electricity was largely privatised after 1998. The results of privatisation have not been popular.[65] However, hydro is still owned by the state governments in Quebec and British Columbia.

In Alberta, electricity generation is mainly fossil fuel based and owned privately. In 2023 the government of Alberta even imposed a moratorium on the development of large wind and solar power projects, seemingly believing that such renewable energy projects were a greater threat to the environment than the state's massive tar sands industry. Of course, the bulk of all energy generated and supplied in Canada (not just electricity, which is predominantly renewable) is oil gas and coal, and the companies that own the bulk of these assets are big private corporations. Overall, the privately owned energy corporations are slowing down decarbonisation since they are mainly involved in building new fossil fuel assets rather than renewables or energy efficiency.

The federal government wants a 'net zero' electricity system to be in place in 2035. It is proving to be a challenge to persuade provincial governments such as Ontario's, where the Progressive Conservatives have governed the most populated province since 2018. Jack Gibbons, of the Ontario Clean Air Alliance said:

The Ontario Government has a procurement out for up to 1,500 MW of new gas power plants. In 2017 gas provided us with only 4% of our electricity for the provincial grid. Last year it was 10%

and this year (2023) it is going to be 14%. By 2043 it will be 27% so we are definitely going in the wrong direction. We are asking the Federal Government to bring in regulations to ban the building of new gas plants in Ontario effective immediately and to move the Ontario electricity grid to net zero by 2030.[66]

The Ontario government is also supporting the extension of heating using natural gas. Unfortunately, heat pumps, which will use electricity three or more times as efficiently as conventional supply, are not being given preferred status in providing heating needs. Despite these limitations to its strategy to reduce carbon emissions, the Ontario government has indicated support for a new nuclear power programme based around SMRs.

The Ontario government will fund the new nuclear programme out of consumers' pockets. Going on the basis of other attempts to build nuclear power in the West, this could increase consumer bills to such an extent that the nuclear building programme will be foreshortened. As usual with nuclear power, the costs are likely to greatly exceed what the hopeful supporters say at the time. Yet the Ontario government has sidelined much cheaper wind and solar technologies.

Quebec is the second largest province in Canada, and 94 per cent of electricity comes from hydro resources, and most of the rest from wind power.[67] It is developing new wind farms. Quebec's electricity, as sold to consumers, is by far the cheapest in Canada.[68] Most of the electricity generation is owned, distributed and supplied by the publicly owned company Hydro-Québec. In 2017, Quebec formulated a plan to phase out coal use and also to develop a large energy efficiency programme. Quebec has a progressive policy on heating services. Installing fossil fuel heating in new buildings was banned from the end of 2021 and banned in existing buildings from the end of 2023. However, most of Canada is not following Quebec's lead in banning fossil fuel boilers, with notable exceptions such as the city of Vancouver.

CONCLUSION

As much as the USA and Canadian federal governments are spurring incentives for green energy, through the efforts of Biden and (to a lesser extent) Trudeau, the path to an energy transition is still slowed by the opposition of energy corporations and their fossil fuel friends. Set against this, global prices and technological pressures are likely to accelerate the shift towards green energy. Natural gas prices are likely to increase and renewable energy costs are falling.

Both Canada and the USA share some key characteristics. First, they are countries with dispersed settlements, which itself encourages more energy consumption. Second, they are leading fossil fuel energy producers, with the USA leading on both oil and natural gas production. This involves powerful political pressures against decarbonisation. Yet there are many who realise that the future of energy is renewable and energy efficient.

In the USA there is a great struggle to decarbonise the electricity market. Canada is much farther down the road on this than the USA, on account of its very large hydroelectric production. However, the fact that Canada is so carbon intensive as an economy highlights the need to pay much greater attention to demand reduction and electrification of the economy.

Canada has a fuel-intensive industrial sector, and too few provincial and city governments are taking the electrification route for providing heat in buildings. Indeed, in some parts of Canada, notably the biggest province of Ontario, gas consumption is actually increasing in both the electricity and the heating sectors. Alberta favours environmentally destructive tar sands but wants to restrict wind and solar power. On the other hand, Quebec is much more oriented towards green energy objectives in terms of energy efficiency and renewable energy supply. Quebec's electricity infrastructure is still in state hands.

The electrification of demand should produce much greater demand for renewable energy. There is a slow growth in use of EVs so far, although there is no attempt to regulate vehicles at a federal level in order to phase out petroleum-based vehicles. The efforts to fully electrify home energy consumption, through switching to heat pumps and electric cooking to replace gas appliances, are piecemeal. Again, there is no federal programme to achieve this, and there is no federal programme to guarantee that PPAs are available for renewable energy developers. The USA badly needs to play catch-up with the rest of the world, because otherwise it will find it is failing to compete in today's and tomorrow's green energy technology markets, let alone do its fair share in reducing environmental problems.

The USA's IRA looks like it is effective in acting as a magnet for companies making electric cars and batteries. It also stabilises the incentives that have been available for renewable energy in the past, and promotes EVs and heat pumps. However, what is missing is any guarantee that renewable energy projects will be offered PPAs, enabling them to go ahead. There is also no guarantee that state policies will favour a shift away from using gas stoves and heating appliances and towards heat pumps. There is insufficient support for electric cars and even hostility towards them in some states. This situation must be reversed. PPAs must be guaranteed, and rules and incentives are needed in order to electrify heating and transport. There is a danger that fossil fuel interests may be given substantial funds to engage in demonstration programmes involving fossil fuel power plant CCS and infrastructure that does not lead to sustainable or even practical, economic outcomes.

It is easy to blame Republican politicians in holding up green energy progress – some Republicans are better than others on this issue. However, if it was not for the strength of the fossil fuel interests and the utilities that use them, there would be much greater toleration by conservatives for clean energy. Indeed, conservatives in Europe are often keener about professing their support for green energy.

What is certain is that, regardless of party-political dominance, the same general pro-corporate/utility malaise has pervaded the US energy regulatory system. This includes the bias against independent companies attempting to develop green energy projects – in part simply because existing utilities could gain greater returns on building their own power plants, rather than allowing independents to supply electricity. It includes the repression of residential solar installations both directly, through limitations on their financing and taxation levels, and especially seriously through the red tape and delays that face solar installations. Until the implementation of the IRA, there was also the serious issue of bias against publicly owned power companies, community renewable companies and individual households, since they could not usually access the PTC incentives needed to develop renewable energy projects.

Regardless of the existence of federally based tax incentives for renewable energy and other factors favouring it, renewable energy cannot expand quickly enough unless it is assured PPAs. Currently there is no federal means to ensure that long-term PPAs with competitive prices are offered at state or province level. There are also insufficient incentives for infrastructure planners to develop enough power lines quickly enough to develop even the schemes currently in the pipeline. Perhaps even more importantly, decarbonisation will not proceed rapidly unless services (mainly heating) supplied by fossil fuels are themselves electrified through use of heat pumps. Change may well yet come, but at the moment it is not fast enough.

4

The EU: Neoliberalism Has Failed, What Could Be Next?

This chapter examines the EU's policy on natural gas, its response to the recent energy crisis and its policies on the energy transition away from fossil fuels.[1] This is of central importance given the threat to energy security presented by the EU's reliance on imported natural gas, as well as the need to reduce natural gas consumption in order to reduce greenhouse gas emissions. Following on from Chapter 4 on the USA and Canada, I explore how it is that European countries appear to have made more progress on green energy, but how this has been hampered by neoliberalism.

The neoliberal strategy was clearly counterproductive. In the name of extending competition in natural gas markets the EU not only allowed increasing natural gas imports but also discouraged anybody signing long-term contracts stabilising gas prices. I discuss relations with Russia regarding natural gas and, importantly, EU strategies on renewable energy, carbon emissions and carbon pricing through the EU Emissions Trading System. I look, in detail, at the changes in energy policies and practices in some key countries: Denmark, Germany and France. I select Denmark because of its pivotal role in green energy innovation, and France and Germany because of their contrasting approaches to nuclear power and renewable energy.

Unlike the USA, Europe has been far from self-sufficient in natural gas supplies, and so has been deeply affected by recent energy price crises. European gas production peaked in 2004, which then accounted for around half of European consumption. Depletion of gas fields in the UK and the Netherlands saw production decline, so that by 2021 production in Europe was little more than a third of the level of consumption. A comparison between natural gas prices in USA and Germany is shown in Figure 4.1.

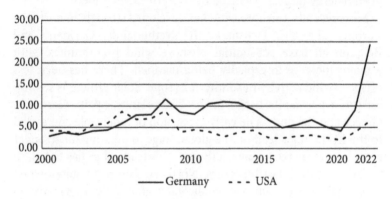

Figure 4.1   Natural gas prices in the USA and Germany since 2000
Source: EISRW 2023.

Campaigns for renewable energy and energy efficiency in Europe began after the oil crisis of 1973, and in many states formed the main response of anti-nuclear movements. The point is that for the renewables industry to grow, it has helped that there is strong state backing and intervention, and that this is likely to be accompanied and driven by strong bottom-up, grassroots movements in support of renewable energy.

This can be seen in the political history of renewables in Germany, Denmark and Spain, where growth in renewable energy has been relatively rapid. This was in the shadow of a national debate about energy futures from the early 1970s and grassroots action to innovate in locally based energy schemes, especially in Denmark where the modern wind industry emerged. Germany's onshore wind movement, and later solar PV movement, was spurred on by independent, cooperative and municipally backed initiatives. Spain's renewable energy industry began with the Spanish state sponsoring technical development and public–private partnerships with wind power developers. Feed-in tariffs were, again, a crucial part of the expansion of renewable energy. All three countries were driven partly by an aversion to the idea of nuclear power as a solution, and such an aim inspired many grassroots renewable energy activists. Eastern European countries who joined the EU in 2004 have generally been slower to adapt to the green energy targets advanced by western EU members.

More recently, offshore wind has developed strongly in northern Europe. The UK, Denmark, the Netherlands, Germany and Belgium all have substantial offshore wind programmes, with a major increase in capacity being planned. There has been less offshore wind activity in southern Europe. Solar growth is strongest (on a per capita basis) in the Netherlands, Spain, Germany, Greece and Belgium. The early expansion of renewable energy led to the EU taking a lead on the global stage in pressing for stronger greenhouse gas reductions, although this leadership has faded, to an extent, in the last ten years. Northern European state-owned companies (e.g. Orsted, Vattenfall, Stat kraft and Equinor) have, to a greater or lesser extent, been very active in deploying offshore as well as onshore wind.

Often the big energy corporations opposed the incentives offered to renewable energy. However, political support for renewable energy has proved too strong to prevent rapid development of renewables, so ultimately the big energy corporations (outside France at least) have tended to join in with renewable energy as the principal future mode of development. In some countries, notably Spain, the big energy corporations were the main means of renewable energy development from the early 1990s. In the UK, where there was much less grassroots action in promoting schemes for renewables, the design of the incentive system protected the big energy companies from losing money (as mentioned in Chapter 2).

NEOLIBERALISM AND ENERGY SECURITY

The faults with the EU's neoliberal strategy were exposed by the natural gas price crisis and the Russian invasion of Ukraine. If, instead of encouraging competition in a globalised natural gas market, the EU had pursued a strategy of reducing reliance on natural gas use in general, then Europe would have been in a much better position to deal with the energy security crisis and reduce greenhouse gas emissions. Neoliberalism, energy security and climate protection are clearly not a good match!

So how did liberalisation emerge? Part of the reason for the existence of the EU has been the building of a seamless market, where competition would (it was hoped) bring European citizens the fruits of goods and services of high quality at low cost. As the 1990s progressed, EU legislation stressed the creation of competitive liberalised energy markets, first inside EU countries. This was influenced by the UK's privatisation and liberalisation programme, so it is a paradox that the UK left the EU. In the early twenty-first century the attention of EU leaders shifted to the creation of a single EU energy market. This involved more rafts of EU legislation, promoting the harmonisation of energy markets. The objective of more competition in energy markets was given priority over climate objectives and, it turned out, long-term security.

Eastern European countries were particularly vulnerable to Russian demands for them to pay high prices for their gas imports. This was simply because if eastern European countries wanted to buy natural gas they had little alternative than to buy it from Russian sources.[2] Russian companies tended to agree contract prices based on whether countries could buy gas at cheaper prices somewhere else (e.g. from Norway or the Netherlands) with concessions apparently made to places that were politically close to Russia.

A liberalised market would, it was argued, protect against monopolistic practices said to be employed by the Russian companies such as restrictions on whether gas could be resold to third parties. The European Commission (EC) promoted a policy of abandoning long-term contracts for gas supply. More emphasis would be placed on global spot market prices. The global gas market was supplied with LNG, often sourced from the Middle East. In the first two decades of the century, global gas prices tended to be low, as LNG supply usually exceeded demand. This meant that European countries could save money by purchasing on the spot markets rather than being tied to prices agreed according to long-term contracts. However, world LNG prices can, and do, go up as well as down. In 2021 they went up.

Earlier in the century, eastern European countries failed to persuade the EU to set up a common bargaining front with Russia for gas supplies. Germany thought it could do better by bilateral deals with Russia. This was done through the Nordstream 1 gas pipeline (ordered in 2005, completed in 2011), which supplied gas directly from Russia to Germany across the Baltic Sea. Later, plans for a Nordstream 2 project were developed, also across the Baltic Sea, which was meant to further increase direct supplies from Russia. Indeed, the former Social Democrat chancellor, Gerhard Schroder, who supported the development of Nordstream 1, was employed to lead the development of the pipeline. Nordstream 2 never quite went to work, and Nordstream 1 was abandoned when the Ukraine war started. The pipeline was sabotaged in mysterious circumstances.

Confidence in Russian gas supplies had already been shaken when disputes between Russia and Ukraine over payments for gas supplies seriously disrupted European gas supplies in 2006 and 2009. Bulgarians in particular were left very cold in 2006. Eastern European suspicions of Russia were exacerbated when, in 2014, Russia seized Crimea and supported pro-Russian separatists in eastern Ukraine. Hence there were both geopolitical and ecological arguments that Germany should reduce reliance on Russian gas and natural gas in general. Eastern European countries opposed the building of Nordstream 2, as they would have received less income from gas than they would if it was transported through pipelines going across their own countries. Then came the Russian invasion of Ukraine on 24 February 2022.

SPIKING NATURAL GAS PRICES

It is not clear if the Russian government thought that Germany and other gas clients, such as Italy, would continue buying gas more or less as usual after the invasion of Ukraine. Perhaps, ultimately, Putin judged that the European gas supply business was expendable anyway, in the face of Russian nationalist objectives. Russian gas exports to Europe were mainly transported through pipelines, and such gas production cannot easily, if at all, be repurposed to be sent anywhere else. Nevertheless, the result was the run-down of the business of supplying Russian gas to Europe, as well as imposing a price limit on imports of Russian oil.

The Russian invasion of Ukraine produced a major shake-up in EU gas policy. There were now two key imperatives. One was to deal with the gas price shock. The other was the political objective of sanctioning Russia and ending reliance on imports of Russian gas. The implicit policy, of allowing Germany to increase gas imports from Russia, was abandoned. Rather the EC and its member states (who do most of the policy work) sought to replace the natural gas coming from Russia, partly by shifting gas imports to other sources and partly by boosting production of green hydrogen.

The EU came up with a plan to end its reliance on Russian gas through a combination of extending renewable energy, energy efficiency, obtaining natural gas from elsewhere and producing large quantities of green hydrogen from renewable energy. The emphasis on the extra production of green hydrogen is a very dubious objective, since it is an inefficient way to use renewable energy compared to using it in heat pumps. The EU is still being driven down the wrong road by big energy corporations who favour policies that help themselves but not the public at large. The German think tank Agora Energiewende has argued that the EU does not need all this hydrogen and that more emphasis could be placed on the heat network, additional renewable energy solutions such as geothermal energy and the electrification of industrial processes.[3]

It may be simplistic to see the European gas crisis as being solely concerned with the Russian invasion of Ukraine. The crisis began to emerge with increasing global prices of LNG as early as May 2021,[4] well before the Russian invasion of Ukraine. Prior to the invasion of Ukraine, the Russian strategy of limiting exports of natural gas to Europe may well have been influenced by Russian military strategy. The point is that they would not have been able to do this but for the fact that the EU had become hostage to a reliance on a tightening world LNG market as an alternative to Russian supplies.

The problem was that by the beginning of the 2020s, the LNG market had ceased to be a buyer's market. Rather, with increasing demand from the East (from countries such as China and South Korea), the global gas market had become a seller's market. China has been building up its natural gas use as a means of replacing the coal used in heating services. In 2021 there came a point where global supply could no longer meet this increased demand so easily. US prices remained well below other prices. This is because the large bulk of trade in natural gas is still through pipelines rather than being shipped, as is the case with oil. Hence regions that have self-sufficiency in natural gas (such as North America) will be subject to much less pressure from price increases in the global LNG market.

Although Russian companies such as Gazprom were previously selling to Europe for relatively low prices, now they could charge more, because the Europeans would otherwise have to buy from the global (shipped) LNG market at much increased prices. As such, European consumers (in late 2021) paid much higher prices for their natural gas sent through pipelines from Siberia in Russia. In earlier years this situation would not have emerged so suddenly, since prices were fixed through long-term contracts. However, the majority of these had not been renewed (in line with neoliberal EU policy) and Europe was mostly exposed to spot market prices. In other words, neoliberal ideology was to blame for the severity of the European gas crisis.

The worst case of this free market ideology was to be seen in the case of the UK. Not only did they not have any long-term natural gas contracts, but they did not encourage the building of any gas storage facilities. Energy security was left to the market.

Despite this experience, there is no explicit alteration of the EU's commitment to neoliberal theory. But its practice has been altered, at least by the fact that the EU plans to phase out use of Russian gas as an energy security priority. This gives a major boost to renewable energy and demand-side reduction measures such as heat pumps.

However, EU policy is still giving succour to the big fossil fuel interests in pursuit of ineffective and costly hydrogen and carbon capture technologies that are still based on fossil fuels. So-called 'blue hydrogen' derived from natural gas is something that favours existing oil and gas interests. As discussed in Chapter 1, it is much less efficient as a solution compared to demand reduction techniques and electrification using renewables. Further evidence for the influence of corporate interests in slowing EU energy transition is seen in the fact that the EU caved in to fossil fuel-addicted automotive car interests. The EU has agreed to allow the use of 'synthetic' fuels after 2035 to power cars. Such fuels have very low fuel efficiency compared to electric cars.

It is important to focus attention on how the overconsumption of natural gas was fundamental to the European gas supply crisis rather than the (Russian) source of the gas. Similarly, in terms of solutions, it is important to focus the biggest efforts on reducing the need for natural gas, rather than simply reacting to the supply side calls for more LNG handling facilities or more pipeline deals. This should make obvious sense in terms of energy security, since not only are other supply sources inherently open to disruption but also the alternative supplies still emit greenhouse gases.

EU AND GREEN ENERGY

The EU promulgates many energy policies but only some have mandatory, direct effects as regulations or directives. The Renewable Energy Directive of 2009 may be regarded as especially influential. Overall, it set a target of achieving 20 per cent of final energy delivered by renewables by 2020 with individual mandatory targets for each state. The revised target, set in 2023, was for the achievement of 45 per cent of energy to be supplied from renewables by 2030.[5] However, there are no individual member state targets, which greatly reduces the political impact of the directive.

The 2009 Renewable Energy Directive was very important, since even before it was finally promulgated, it enabled NGOs and renewable trade groups to pressure governments for action to promote renewables.[6] The 2009 directive emerged not only after considerable activism against nuclear power and in support of renewable energy in states such as Germany, Spain and Denmark[7] but also in the wake of an increase in energy prices from 2005 onwards. However, the 2009 directive gave prominence to the use of biomass energy as well as wind and solar power. Biomass gave some eastern European countries greater possibilities for reaching their renewable energy targets. The sustainability of biomass resources became an item of controversy.

The EU was in the vanguard of developing renewable energy from an industrial viewpoint, since in the early 2000s, wind and solar industries were largely based in Europe. This lead waned, though, as China steadily expanded its production of solar PV, utilising the very markets that the Europeans created to capture economies of scale which quickly brought down costs. Some conservatives in Europe complained about the cost of incentives going to solar PV in particular. Hence the initial scale of the feed-in tariffs for solar PV were reduced. Despite the complaints about initial costs, a mass market for the technology was created, which resulted in great cost reductions. Overall the strategy was a tremendous success.

The EU retains nominal control over the nature of nationally based incentive schemes for renewable energy. However, in practice the details are left up to nation states. The dominant method up to around 2010 was feed-in tariffs, wherein generators are paid a pre-set, guaranteed price for renewable energy over a long-term period. After 2010, at the level of larger projects, such contracts have been awarded as a result of competitive tendering.

There has also been an increase in the use of CfD systems. A difference compared with feed-in tariffs is that these contracts are awarded through auctions, with developers offering the lowest prices being awarded contracts. These contracts are like feed-in tariffs in that generators are guaranteed fixed prices. The generators are responsible for selling their power onto power markets, and then a state agency refunds (or deducts) the income that they receive to bring payments in line with their guaranteed price. This leaves out smaller, community-based projects that need the old system of fixed prices set by administrative means. The UK and Spain used to use a market-based renewable energy support system, wherein renewable energy incentives were supplementary to the income that generators could gain from the sale of their energy on wholesale power markets. These countries have now switched to fixed price contracts, after it became clear that the market-based systems produced renewable energy at rather greater cost.

Europe has seen programmes that have involved high (initial) seed corn funding for specific technologies, whether it be onshore/offshore wind or solar PV. For example, in several European countries, early funding was very generous for offshore wind farms, with initial contracts offering high prices for the energy produced. The processes for issuing long-term contracts for renewable energy in countries such as the UK, Germany, Denmark, the Netherlands and Belgium have been well organised by the state and are notable for their relative consistency. This has allowed costs to come down in recent years and for the volume of contracts issued to be greatly expanded.

Emerging green technologies have been funded by the European Investment Bank (EIB). The EU took the controversial decision in 2022 of declaring that nuclear power and gas-fired power plants could both qualify as green technologies. This raises the spectre that EIB investments in new green energy technologies such as floating wind and rural renewable energy systems in Africa could be crowded out by massive lossmaking investments in new nuclear power plants.

The EU has introduced several measures, and promulgated a series of directives, aimed at improving energy efficiency. However, the most direct instruments have probably been on standards for appliances. Laws banning use of tungsten lightbulbs were passed in 2009, and these have had a substantial effect in assuring an early move to fluorescent bulbs and now LEDs – certainly, progress on this happened earlier than in the USA.

The EU also introduced minimum energy efficiency standards for appliances. On top of this, energy labelling was introduced not only for energy appliances but also for buildings. Heat pumps are being promoted vigorously in most European countries, although this is being organised at a state rather than EU level. The EU mandated motor vehicle manufacturers to achieve average fleet fuel efficiency improvements.

In later years, campaigns for people to install heat pumps have gathered pace, with Italy offering perhaps the largest incentives for heat pumps in 2022. Northern Europe (excepting the UK) has seen a big growth in the use of heat pumps, and this growth includes Finland, which attests to their suitability for very cold climates. The growth in EU heat pump installations can be seen in Figure 4.2.

In 2019, the EU agreed to adopt a target of achieving net zero greenhouse gas emissions by 2050. Eastern European states, notably Poland, Hungary and the Czech Republic, were reluctant to agree to the target.[8] Poland initially opted out of the agreement, although it joined the following year, given the offer of substan-

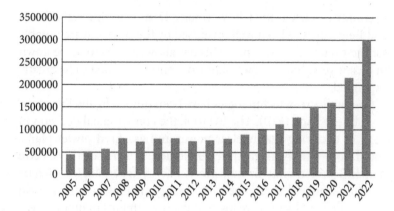

Figure 4.2   Heat pump installations in the EU, 2005–22
Source: European Heat Pump Association; European Heat Pump Market and Statics
Report 2023.

tial funds to help achieve carbon emissions reductions. Poland has
been increasing its energy use (5 per cent over the decade from
2011 to 2021) and nearly half of Poland's energy came from coal
in 2021.

THE EU EMISSIONS TRADING SCHEME

The EU hoped to reduce carbon emissions with the adoption of
the EU Emissions Trading Scheme (ETS). This was the first such
scheme in the world applied on a large scale. Ironically it was sug-
gested by the US Clinton administration and initially resisted by
the EU as a means of implementing the Kyoto Protocol, which was
agreed in 1997. However, the USA never actually implemented
an ETS. The EU was left 'holding the baby'. The idea behind
the system is that the EU sets a ceiling that defines the allowa-
ble carbon emissions for the whole of the EU. In order to burn
fossil fuels, companies have to, in theory, pay money for emissions
allowances which were issued initially to companies on the basis of
their historical emissions. Hence, according to the theory, companies which find it relatively easy to reduce emissions can sell their
allowances, while companies that find it more expensive to reduce
emissions will buy the allowances.

The EU started the scheme in 2005, but it has never lived up to expectations. One big factor is that the 'ceiling' for emissions, covering the whole of the EU, was always quite high, on the insistence of fossil fuel users who feared their costs might radically increase. The tendency has been for the emissions ceiling to be too high – below what emissions might have been in a business-as-usual situation – rendering the policy limited in effect. An additional issue is simply that the sheer uncertainty about the level of future prices of emissions allowances meant that there was little incentive from the ETS to invest in energy assets, such as energy efficient equipment or non-fossil energy technologies. The EU ETS did have an effect of hastening the shift from coal to gas generation. But it could not have any significant effect on promoting renewable energy. This is because practically all renewable energy generation relies on incentives (usually contracts) guaranteeing prices for generated outputs set by the member states.

Investors in new renewable schemes are going to rely on guaranteed payments or incentives for generation rather than highly uncertain possibilities of future levels of emissions allowance prices. Investments in energy efficiency investments are guided mainly by regulations and incentives. For that purpose, the EU ETS is mostly irrelevant. A careful EU-funded academic study of the impact of the EU ETS found that, outside of the switching between use of different conventional power plant, the effects seemed modest and difficult to distinguish from other overlapping regulatory measures.[9] The EU ETS is largely a testament to how neoliberal notions have distracted attention away from direct measures to encourage renewable energy and energy efficiency.

A more general criticism of the reliance on the EU ETS, urged by many supporters of neoliberal methods, is that this misconstrues the necessary objectives of the energy transition. The energy transition must be about adopting new technologies which are thought to be suitable on policy grounds. Carbon taxes or carbon pricing may sound good, but in reality they do little to make corporations adopt technical change other than switching between technologies that they are already using.

Much the same can be said of the wider neoliberal agenda organised by the EC in the energy sector. What has been effective, for example, has not been generalised, liberalised, energy markets, but good regulation to support energy efficiency and good support for renewable energy generators. The latter includes the ability of independent green energy companies to compete to generate energy in a market that is in effect separated and protected from the fossil fuel market. This has to be achieved at the expense of the fossil fuel-controlled energy corporations, and moreover, now that renewable energy is cheaper than fossil fuels, the savings in switching to renewables should be passed through to consumers. The growth of renewable energy had nothing to do with (in fact it mostly preceded) the turn towards liberalising energy markets, which gathered pace in the EU at the beginning of this century. It is this emergence which has flourished and threatened to replace the prevailing energy regime.

The EU has attempted to reform energy markets, but Jakob Embacher and Stephen Thomas have described such attempts as a 'sticking plaster' and called for much greater direct state intervention.[10] The European Federation of Public Service Unions (EFPSU) has said much the same thing as I have in this section.[11] They criticised the EU's energy market liberalisation as a failure (see the discussion of the British case in Chapter 2). The EFPSU has called for natural monopolies such as supply and electricity grid and distribution to be taken back into public ownership.

DENMARK

I go to Denmark first, because it exhibited the most radical direction and inception of revolutionary changes towards green energy, at least at the level of energy generation technologies. In Chapter 1, I summarised the Danish grassroots energy revolution as forming the basis for the modern wind energy industry. From the end of the 1970s lots of wind turbine manufacturers sprang up, initially at local blacksmiths, and there was a movement of cooperatives and farmers putting up wind turbines. The Danish cooperative tradition was an important factor in this process.

The farmers, who were the first to install wind turbines, demanded that they be given payments for excess power sent into the electricity grid. This was the start of the term 'feed-in tariff'.

Besides farmer-owned turbines, cooperative wind associations grew up, owning increasingly large numbers of wind projects. Shares in the cooperatives were funded through a form of personal tax incentive, whereby wind co-op shareowners received rebates from the energy taxes that they would normally pay.

This was only part of the energy revolution. District heating systems sprung up all over Denmark, encouraged by tax incentives organised by the national government. These district heating systems were powered by small, gas-fired combined heat and power plants and existing coal-fired power plants. The most successful programme of heat networks in the world appears to be in Denmark. Nearly 70 per cent of the heat supplied in the country comes from district heating networks which are owned either by municipal authorities or local cooperatives.[12]

More recently, large heat pumps have been installed to replace the coal-fired power plants, which have been supplying heat to some municipal networks. The installation of fossil fuel boilers in new buildings was banned from 2013.[13]

Denmark aims to provide 100 per cent of its electricity from renewable energy by 2030. Denmark is already well over halfway to its target, with the bulk of renewable energy coming from wind power. Denmark had the highest quantity of wind and solar power generation per person of any country in the world in 2022, with Norway and Sweden in second and third place.[14]

Since the beginning of the century, the main increase in wind power capacity has come from a growing capacity of offshore wind schemes, which have, in the last decade, continued to become cheaper in terms of the price paid for the electricity generated. The Danish government is increasing the pipeline of offshore wind projects. Danish energy consumption fell by 15.2 per cent in 2021 compared to 2011.[15] The build-up of renewable energy can be seen in Figure 4.3. Electricity from fossil fuels has mainly been eliminated, with wind and solar (mostly wind) making up 61 per cent of electricity generation in 2022.

Danes cycle much more than most other countries. Around nine out of ten Danes own a bicycle, but fewer than half of them own a motor vehicle.[16] I recall working in Aalborg in Denmark in

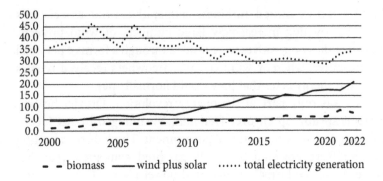

Figure 4.3 Electricity generation in Denmark from wind, solar, biomass and fossil fuels in TWh per year
Source: EISRW 2023.

2003 for three weeks writing an (ultimately successful) EU funding proposal at the city's university. I thought the pay I was being given was very high, until I realised that the level of taxes was very high compared to the UK, where I normally work. On the other hand, Denmark's social welfare system is a comprehensive one, as is cooperation between the state and trade unions, something that is noticably absent also from the UK by comparison. Perhaps this is related to the feeling of social responsibility and support for green objectives. Nevertheless, Denmark has a very high standard of living and a per capita GDP that is higher than the UK.

The Danish energy industry, like the rest of the EU, underwent privatisation and liberalisation measures in the early years of this century. However, the government still maintained a majority share in its oil and gas company, DONG. In 2017, DONG sold off its hydrocarbon assets and became a solely renewable energy developer with a new name: Orsted.

Orsted competes for big offshore wind contracts with other Nordic state-owned companies from Sweden and Norway. Denmark's Vestas is the largest wind turbine manufacturer in the world. The Danish renewable energy industry is not 'small' anymore, but to become big, it makes a big difference how things are started. Bottom-up revolutions can make an enormous difference. They do so because the same well of national impetus for sustainable energy is needed to sustain support for large-scale development. The weight of inertia and the accumulation of skills has propelled Denmark forwards, especially in the wind industry.

GERMANY

Germany had a broadly similar green energy trajectory to Denmark from the start. This spawned independent wind developers who challenged the main utilities. The political force of the movement led the government to set ambitious targets for renewable energy development. A system of feed-in tariffs was established, first for wind power and then for solar PV.

In terms of political economy, Germany has championed the 'social market economy', mixing up an emphasis on private ownership and social welfare policies. Along with EU policies of liberalisation, there has been an effort to curb monopolies in the energy sector and introduce competition through liberalisation. However, it should also be recalled that German policy, when compared to the UK, appears to involve more energy planning. Germany has gas storage capacity equivalent to around a quarter of annual national consumption, while the UK has practically none. Indeed, the gas storage held by Germany and other EU states was essential in avoiding a much worse European energy crisis in 2022.

The movement for renewable energy began in the 1980s, as an offshoot of the anti-nuclear movement. Unlike Denmark, Germany built up a major nuclear power industry based on German pride as leading engineers. However, given Germany's experience with authoritarianism, many on the left became suspicious of the nuclear energy industry, seeing it as a centralised authoritarian threat, and anti-nuclear politics overlapped with movements against stationing nuclear cruise missiles in Germany. In addition, in the 1980s there was the phenomenon of local 'citizens' initiatives' being formed to act on a range of environmental issues. Die Grunen, the Germany Green Party, formed because of a confluence of these political phenomena. They gained strength in the 1980s, and the Chernobyl accident in 1986 was interpreted very strongly in an anti-nuclear direction.[17] After the Fukushima accident in 2011, the Christian Democrat-led government, under serious electoral pressure from the Greens, announced a policy of phasing out nuclear power.

Many anti-nuclear activists decided to support alternative energy paths, and organising commercial wind farms was part of this. From 1990, following lobbying by independent companies and farmers wishing to start up renewable energy schemes (mainly wind power), the German feed-in tariff came into action. Initially, this was mainly concerned with wind power, but a feed-in tariff was established for solar PV in 2000. Independent wind power companies sold shareholdings to affluent Germans who could offset the investments against tax. The bulk of the capital needed to establish the projects was borrowed from banks who were assured of the money being repaid because of the security of the government-backed feed-in tariff.

The campaigns for the feed-in tariffs were supported by a wide range of grassroots citizen organisations. Cities even introduced their own feed-in tariffs, which led to the adoption of a national feed-in tariff system for solar PV. Germany was an early adopter of solar feed-in tariffs and a big market for solar PV was thus created. It has been criticised as being a very expensive programme – one that was/is paid for by a levy on consumer energy bills. Nevertheless, the German solar PV market, and similar schemes in other European countries that emerged later, created a much-enhanced market that led to a steep decline in the cost of solar PV. Unfortunately for German solar PV producers, Chinese manufacturers then stepped in to dominate the solar PV manufacturing market. Indeed, for a time, German solar PV interests managed to argue successfully for EU restrictions on imports of solar panels from China.

The energy utilities, which were privately owned and vertically integrated companies controlling generation, transmission/distribution and supply, were fiercely opposed to the feed-in tariff incentives given to renewable energy companies. Until the inception of the offshore wind programme, the feed-in tariffs existed solely for the benefit of the independent renewable energy movement. The utilities campaigned against the feed-in tariffs by saying that wind farms were bad value for money for the consumer.

The utilities wanted to defend their coal-fired power plants from being undermined. In some ways there are parallels here between the way US utilities have fought the incentives given to renewables that competed with their own power plant outputs. But a big difference is that the political support for the German independent renewable generators was, and is, very strong. Hence the independent renewable generators were given good contracts guaranteeing payments for energy outputs.

However, this political competition did have the effect that the feed-in tariffs were pared down to pay the developers what was needed to organise the projects. Ironically this meant that feed-in tariffs for wind power in Germany were lower than the incentives being paid for wind farms in the UK.[18] This is despite the fact that the wind farms built in the UK were usually built on much windier sites. The political difference was that, following the launch of the UK Renewables Obligation in 2002, most of the wind farms were bought up by the big energy corporations. They had much less interest in trying to argue for lower incentives for renewable energy schemes, since they were beneficiaries of the schemes. Indeed, I recall a German wind power official who I encountered at a conference in 2012 exclaiming to me (as a British activist), 'How do you manage to make wind power so expensive?!'

While renewable energy has continued to expand rapidly between 2011 and 2021, oil and gas consumption has also increased. Coal has declined, as has nuclear, which by 2021 was well on the way towards being phased out. German primary energy consumption fell by 5.4 per cent in this period, which is less than some other western European countries. The Russian invasion of Ukraine convinced the German government to end their strategy of importing gas from Russia. Gas imports from Russia declined by a factor of almost three in 2022 compared to 2021, and this looks set to decline even further in the future.[19] Germany has been building new LNG handling facilities so that supplies of natural gas can be shipped in from the USA and the Middle East.

The German Greens, who were now in the government coalition, were among the strongest supporters of Ukraine within Germany. They had also argued against Nordstream 2 on ecological as well as economic grounds.[20] In energy policy terms at least, there was congruence between support for Ukraine and reducing gas use. After all, the Russians promoted reliance on natural gas and nuclear power, two fuels that the Greens wanted to phase out in favour of renewable energy and energy efficiency.

The entry of the Greens into the government coalition in 2021 saw the renewable energy target increased to 80 per cent of electricity generation in 2030 and 100 per cent by 2035; 47 per cent of electricity was generated by renewables in 2022.[21] Wind power installations need to increase sharply to enable this target to be achieved, and planning opposition, which has seen approvals slow down, needs to be overcome for targets to be achieved. The Green energy and economy minister, Robert Habeck, formulated laws that impel state governments to speed up planning approvals for wind power.[22]

The government also planned a large increase in the rate of installing solar farms and solar PV on buildings. Previously, solar energy development on farmland was discouraged, but this has been reversed and solar farms which involve 'agrivoltaics' will be incentivised. Certainly, there will be sufficient fixed price contracts available for renewable energy generation. The only issue is whether planning permission is given to enough schemes to fulfil the rate of increase in renewables necessary to meet the target for renewables expansion.

Germany continued the phase-out of nuclear power even during the natural gas crisis. Here there is a paradox. Many argue that the drive for an early phase-out of nuclear power is against a carbon reduction strategy. Yet the political dynamic in Germany is that much of the drive for renewable energy has come from anti-nuclear aspirations. The comparison can be seen in Figure 4.4., and with nuclear production run-down in France in Figure 4.6.

More renewables have come online compared to the amount of nuclear power that has gone offline. However, in 2021–2 there was an increase in coal use. This was caused by the increase in prices of power from natural gas and (consequently) a decline in the use of power from natural gas plants as well as the nuclear phase-out. Then, in the first part of 2023, renewable energy increased rapidly and coal production declined. Clearly, renewable energy installation needs to dramatically accelerate in the coming years. Changes

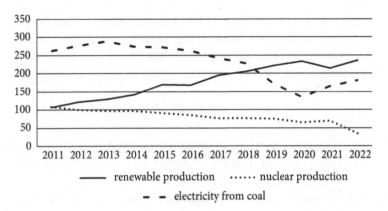

Figure 4.4 German electricity production from non-hydro renewables, nuclear and coal, 2011–22 in TWh per year

Source: EISRW 2023.

in electricity production from renewables, nuclear and coal can be seen in Figure 4.4

Controversy has raged about the extent to which gas should be completely phased out and the extent to which some should still be used as a feedstock for the production of hydrogen and synthetic fuels for motor vehicles. Green pressure groups have argued against the use of hydrogen in heating. Fossil fuel boilers are being banned in new buildings from 2025 in favour of heat pumps, whose use in Germany has taken off in recent years. However, the German coalition was divided when it came to debating a heating law about phasing out gas boilers in existing buildings. As part of a compromise, municipal authorities have been given the task of making plans for heat networks to be powered by large-scale heat pumps. German car manufacturers have dragged their feet on pressures to ban fossil fuel vehicles from 2035.

A likely weak point in Germany's net zero strategy is the government plan for widespread use of hydrogen in the transport, heating and industrial sectors. A lot of this demand could be covered much more efficiently using renewable electricity rather than hydrogen, which wastes a lot of energy in its production.

An interesting shift away from a total rejection of state ownership of energy assets occurred when the energy crisis hit home in 2022. This involved the German government nationalising two companies. These were Uniper, the major importer of natural gas, which was heading for bankruptcy, as well as a trading arm of Gazprom, the Russian gas company. Post-World War II Germany has involved reliance on private ownership as a state ideology.

FRANCE

France's energy policy has, since the 1970s, been characterised by the emphasis on its nuclear power programme. However, a notable outcome of the energy crisis of 2022 was that the French nuclear sector seriously underperformed and made the energy crisis worse. French nuclear production has been in decline for several years, as seen in Figure 4.5 where declining French nuclear production can be compared to the increasing German renewable energy production shown in Figure 4.4. It should be noted that German renewable energy production has more than compensated for the decline in nuclear production; the same cannot be said in France where increases in renewable energy have not kept pace with falling nuclear energy production.

French energy policy in the 1960s rested on its postcolonial oil deal with Algeria, which gave French consumers privileged access to oil. However, at the end of the 1960s this relationship broke apart, culminating in the nationalisation of oil assets by Algeria

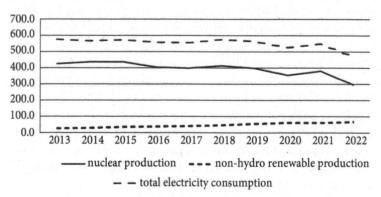

Figure 4.5 French nuclear and renewable production and total electricity production since 2013 in TWh per year

Source: EISRW 2023.

in 1971. This loss of energy security was followed by the global oil crisis of 1973. This created an opening for the nationalised French electricity company EDF to spearhead a nuclear power construction programme. By the 1990s, three-quarters of French electricity was being generated from its nuclear power stations.

In fact, James Jasper wrote that the French Ministry of Finance 'wished to avoid massive borrowing from abroad to finance a technological venture whose benefits did not clearly outweigh the costs'.[23] But, as has happened more than once, the more nuclear-sceptical Finance Ministry was overruled by the pro-nuclear technocracy in favour of EDF. Jasper continues: 'the original enthusiasm for nuclear energy can be traced to the education of the technical elites at the Ecole Polytechnique'.[24]

The centralised nature of the French state, led by a technocratic elite, enabled the French nuclear power programme.[25] This can be compared to Germany, for example, which had less ability to implement policies. Also, the French anti-nuclear protestors had rather more limited recourse to the courts compared to Germany.[26]

There has been little evidence of a bottom-up renewable energy movement in France that is comparable to the renewable energy movements in Denmark and Germany. This is associated with the relative lack of progress in building up renewable energy compared to these two other countries. In other words, more nuclear power may mean less renewable energy.

The weakness of the French anti-nuclear movement (after the 1970s at least) and the muted response to the Chernobyl accident (compared to Germany and Italy)[27] was not the only notable feature of post-Chernobyl French nuclear safety policy. Another was the decision by EDF, initially in collaboration with German nuclear interests, to seek to design a virtually accident proof 'advanced passive design' nuclear power station.[28] What emerged from this was the European Pressurised Reactor (EPR).

In 2005, a minority of shares were floated in EDF, AREVA (the nuclear reactor manufacturer) and the state natural gas company GDF Suez (which later became Engie). These moves were prompted by pressure from the EU for the liberalisation of energy markets.

Local opposition to land-based wind and solar renewable energy schemes has slowed progress. As in the UK, opposition to onshore wind farms is considerable, but differently to the UK, France has found it difficult to quickly develop an offshore wind sector. Opposition from the fisheries industry has been a big problem. France does a little better (compared to the UK) in terms of solar PV. Perhaps the most notable policy innovation concerned with renewables in France has been the adoption of a law, in 2023, making it obligatory for car park developers and operators to install solar PV. Yet France was alone among all 27 EU states in failing to achieve their target for renewable energy for 2020, set under the mandatory targets set out in the 2009 EU Renewable Energy Directive. The EC imposed a €500 million fine on the French government for this failure.[29]

Certainly, France's emphasis on nuclear energy has meant that its relative lack of renewable energy still leaves its energy provision as being more low carbon than most countries. However, nuclear production has fallen substantially and the deployment of new reactors has, so far, failed to materialise. EDF has been building its new EPR at Flamanville since 2007. A series of construction problems and safety controversies has prevented its timely commissioning and has racked up tremendous cost overruns.

The EPR has proved to be a financial disaster in the West at least. A twin EPR scheme was built in China at Taishan, which, though subject to construction delays, was not as disastrous as in the West. AREVA, the French state-owned nuclear constructor, was responsible for building the Olkiluoto EPR in Finland. Construction began in 2005, but the power only started flowing in 2023. Before that AREVA went into insolvency and was restructured by the French government. AREVA had a 'turnkey' contract with a Finnish power group, and although the Finnish contractees received the plant, AREVA lost tremendous amounts of money in the process. The French state effectively paid the price. Something vaguely similar is happening in the UK, where EDF agreed to build a double EPR at Hinkley C for a contract under which the British consumers will pay a controversially high price for the electricity it produces. The plant is still far from built and there are increasing cost overruns in the project. EDF is responsible for the cost overruns. Indeed, in 2015 when EDF decided to go ahead with the project, the chief finance office of EDF resigned as he did not wish to take responsibility for the negative outcome from Hinkley C.

Perhaps the worst news for EDF is that the output from the French nuclear fleet has been steadily declining – French nuclear production fell by 14.5 per cent between 2011 and 2021. However, there is a growing disconnect between French nuclear energy production and total electricity production (see Figure 4.5).

Then things got a lot worse in 2022. Different crises put much of the power plant offline. A hot summer meant that the water supply could not cool some reactors (which had to be taken offline). There was also a general malaise of faults, the discovery of faulty steelwork and a slew of closures due to maintenance calls. Production from reactors dropped by 30 per cent compared to the previous year and EDF posted an enormous financial loss.[30] This breakdown came at the worst possible time, right in the middle of the European energy crisis. The breakdown of French nuclear made power prices shoot up even further than they had been.[31] This was because France was forced to import a lot of energy. This resulted in an increase in demand for power from gas-fired power plants, which exacerbated the crisis. According to a leading energy analyst, Jerome Guillet, talking about high European power prices in 2022, 'gas supply is not the problem for the power sector: the real problem is [the breakdown of] French nuclear, which allows high gas prices to cause power price increases across the board'.[32]

EDF dominates electricity markets in France, and as the French government owned most of the EDF shares, at around 85 per cent, it was able to influence the prices paid by consumers. This was helpful in the management of the energy crisis. The government instructed EDF, in effect, to internalise electricity price rises rather than pass them on to the consumer. This led to large losses for EDF, which has effectively been bailed out by the French government. This level of state control would have benefited French consumers during the crisis much more if it had not been for the crisis in French nuclear power generation.

The French electricity system, long seen as a bastion of nuclear-generated reliability, is facing a crisis because of the steadily declining nuclear output. Not only has EDF suffered mounting losses in France, but the French nuclear state sector has also been dealt a blow through the losses suffered by AREVA in building an EPR in Finland. The result is that in 2023 the French government in effect renationalised the bulk of the electricity industry assets of EDF and AREVA.

A string of financial problems has beset EDF, including pressure from the government to keep electricity prices down. The company also spends considerable amounts of money to keep many of its ailing reactors going. Then there is the issue of new nuclear reactors to be built. Six new nuclear power plants are planned. Hardly any of this can be done on a competitive basis that would leave much profit or value to the shareholders.

The delivery of these six new EPRs is uncertain. It is always curious that, in the West, where there has been such a terrible record of trying to build nuclear reactors this century, people seem to assume that once plans for a new nuclear power plant are announced, this is a done deal, and that it will be ready when the press release says it will start generating. This is a testament to the continuing blindness of the French state technocrats to the realities of nuclear power. There is a harking back to the days of the 1970s and 1980s when the French nuclear fleet was built. But things have changed a lot since then. Mycle Schneider, a French nuclear analyst, has commented:

> The only way to compare what is happening now with what happened in the 1970s and 1980s is to compare early construction in France in the 1980s with China in the 2010s. Like the French then, the Chinese have built up a huge skilled workforce on all levels from the project manager to the welder. They move them from construction site to construction site, so they are getting better all the time. They have a living skills workforce. ...
>
> That is not the case in France today. At the construction site of the EPR Olkiluoto-3 in Finland they have over 50 nationalities in the workforce; this didn't exist in the earlier French construction waves. They were all nationally built. Globalisation has brought multiple problems of communication. They run computers in India for a project in Finland. You have so many more potential possibilities for mistakes and errors in communication. ...

[A]t least one third of the workforce at Flamanville 3 is foreign. The housing and social conditions were so bad – e.g. container flats with no access to public or any other transport – that human rights organizations stepped in at some point. And, of course, globalization is not the only cause for problems (in building the reactors)![33]

Another problem of building nuclear reactors in the present, as opposed to the 1980s, is that the current models of nuclear reactors, such as the EPR, have rather more complex designs, as discussed earlier, which makes them more difficult to build. In addition to this of course, health and safety regulations covering construction of all types have been developed which potentially leads to greater delays and cost overruns. France risks seeing its nuclear power plants running down while not replacing the loss of generation by either increases in nuclear power or enough new renewable energy.

French nuclear production recovered in 2023 to regain its position as a net energy exporter. However, without a much more rapid renewable energy build-up, France will become a long-term net energy importer as existing plants run down.[34] Figure 4.6 shows a comparison between the slow decline in French nuclear power production and the build-up of renewable energy production in Germany.

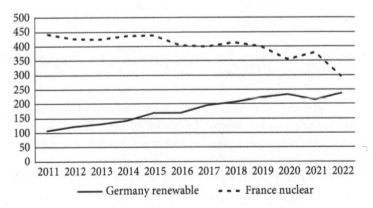

Figure 4.6 Renewable energy production in Germany and nuclear energy production in France, 2011–22 in TWh per year

Source: EISRW 2023.

CONCLUSION

Europe has been an early leader in renewable energy. However, its march towards green objectives has often been waylaid by lobbying from the fossil fuel industries, especially the natural gas industry. The EC planned a neoliberal approach, wherein the best policy for energy security would be natural gas trading via spot markets rather than through long-term contracts. The policy discounted eastern European concerns about reliance on Russian gas and green political pressure to focus on reducing gas use. This neoliberal policy looked shaky when gas prices increased in 2021 and collapsed with the Russian invasion of Ukraine in early 2022.

The EU leadership would have been wiser to focus on reducing natural gas consumption through better state planning. The progress that has been made in reducing fossil fuel use has usually been because of fixed price contracts issued to renewable energy developers and regulations and incentives mandating or encouraging energy efficiency. The market-based EU ETS has only made a modest difference to reducing carbon emissions, and it is questionable whether it has made more than a very small difference in encouraging renewable energy and energy efficiency measures such as heat pumps. On the other hand, it is notable that the activity of state-owned companies from northern Europe has increased competition in the deployment of renewable energy.

Germany and France have charted different pathways towards energy security in that Germany has phased out nuclear power while France has relied heavily on what was once a highly regarded nuclear programme. However, French nuclear power's reputation is declining fast, not least because of its failure to provide security in 2022 – precisely the time when it was most needed. Yet France's dominant state-owned energy corporation, EDF, is still wedded mainly to nuclear power as the key to France's future energy policy. Germany's focus on renewable energy is laudable, but its overenthusiasm for hydrogen seems destined to waste renewable energy that is better supplied through electrification.

It is a testament to some lop-sided debates about Germany and France that while there is much discussion about the German nuclear phase-out there is much less discussion about the recent, rather larger (unintended) phase-out of much French nuclear pro-duction. There is far too little pressure on the French authorities about what is going to be done to increase renewable energy to compensate for the phase-out of French nuclear power. Making promises to build more nuclear power stations is not a sub-stitute for this given the recent history of failure to deliver new nuclear plants.

On top of this, oil and gas interests are pushing for 'blue hydro-gen' and CCS as a key part of energy transition. This is blunting progress towards the adoption of electrification and demand reduction techniques, especially heat pumps.

Europe's progress towards renewable energy and energy effi-ciency objectives has been faster than that of the USA. A key reason for this is that in many European countries, contracts have been issued to pay renewable energy generators guaranteed prices for their power. This emphasis, rather than relying on tax breaks, has allowed renewable energy to displace more fossil fuels. Yet the build-up of renewable energy and energy efficiency is still nowhere near rapid enough. Next we need more planning and interven-tion pushed on by grassroots pressures for policy change. This can turn the corner and revivify the green energy revolution in the EU. The neoliberal approach has abandoned energy security to energy market forces. Rather, a more planned approach based on long-term contracts for energy and more public ownership in sectors such as retail energy supply would have better protected European consumer interests.

5

Getting Green Energy

TECHNOLOGY POLITICS AND ENERGY DEMOCRACY

Getting green energy involves energy democracy. There are said to be 'three dimensions of energy democracy: popular sovereignty, participatory governance, and civic ownership'.[1] This means that informed, ordinary people have the biggest say over technological choices, in terms of deciding policy and using and owning the technologies. That entails policy choices being made by democratically elected governments. It also means that the constellation of forces advising on and implementing policies, and owning technology, must be much broader than the corporations who currently dominate energy policy decision making.

This can be done by people taking community action themselves; by fighting for policies that ensure that the energy industry adopts technologies and practices that meet climate crisis targets; by ensuring that governments implement the climate crisis targets for energy; by ensuring that ordinary people have a chance to influence energy decisions alongside state and private companies; by building strong social movements that give voice to demands for sustainability and equity; by decentralising energy so that, as far as possible, individuals and communities have ownership and control over local energy companies and services.

Getting green energy also means a big role for the state to ensure that green energy works for people to keep down costs rather than giving corporations giant profits. This includes giving municipal authorities more leeway to take action to implement green energy projects. Natural monopolies in energy should be run by the state, if possible, at the municipal/local level. Activity by social movements who are engaging with energy technological choices and practices is also an important part of energy democracy.

Social movements for technological change (what I call 'technology social movements') also include environmental campaign groups. These range from organisations such as the Sierra Club and Greenpeace to more targeted organisations such as Vote Solar (USA) or We Are Possible (UK), who are linked to campaigning for specific technological changes. Associated with these campaigning groups are green energy think tanks such as the Rocky Mountain Institute (USA) and Agora Energiewende (Germany). Brief descriptions of some of the activities of two of these technology social movements (Vote Solar and We Are Possible) are given in Boxes 5.1 and 5.2.

Box 5.1   An example of the US technology social movement

Vote Solar is a non-profit US organisation which campaigns on a state-by-state basis for clean energy, especially solar power. It has a dedicated team of campaigners. In its own words:

> Our team advances clean energy progress in state legislative and regulatory arenas, where most decisions about electricity are made. Since 2002, we have brought our winning combination of deep policy and technical expertise, coalition building, and public engagement to drive meaningful progress. ... Together with partners in local communities, we're expanding solar access, lowering solar costs, and building political power for a just and inclusive clean energy system. We center our work on five programmatic priorities.[2]

Among a long list of policy successes and ongoing campaigns at state level, here are just a couple of examples.

In the Mountain West region, Vote Solar activists in New Mexico led efforts to make participation in community solar available to all and thus to 'remove barriers to energy sovereignty for Tribes and Pueblos'.[3] In Minnesota, Vote Solar campaigners have worked with others to see the passage, in 2023, of a 100 per cent clean energy mandate for the state of Minnesota.[4] In Illinois, Vote Solar partnered with a coalition of other groups to get legislation passed which will ensure:

> Over $80 million per year for solar and energy efficiency workforce development – specifically in Black and Brown communities that have been historically shut out of the clean energy economy. ... Expansion of the Illinois Solar for All programme to provide free solar and guaranteed

savings for low-income families, plus new inclusive finance mechanisms for families to invest in energy updates. ... An end to automatic, rubber-stamped utility rate hikes for consumers and Electric vehicle and transportation incentives.[5]

Box 5.2   An example of the British technology social movement

We Are Possible is a registered charity in the UK and has campaigned on a variety of issues related to cutting greenhouse gas emissions. It was formerly known as 10/10 but was renamed in 2019. Talking about their past achievements, they list the following achievements: '2009–10: The original 10:10 campaign. We rallied over 100,000 people and 7,000 organisations to cut their emissions by 10% in 2010'; 2011–16, Solar Schools, 'Our pioneering campaign worked with 80 schools across the country to crowdfund for bill-busting, carbon-cutting power rooftop solar panels. 2,500+ were installed, saving 40,000 tonnes of carbon – and the model continues to be replicated around the world'; 2013–16, they helped Balcome, 'a sleepy Sussex village', to combat proposals for fracking and instead create a 5MW solar farm'; in 2019 they successfully demanded a TV climate debate among party leaders when they 'lobbied broadcasters and presented a 200,000+ strong petition – and we won! With 1 in 4 people seeing it and 1 in 2 hearing about it, we put climate front and centre amidst the political chaos'.[6]

Most recently, We Are Possible has been campaigning on a range of issues. These include: campaigning for the ending of the ban in England on onshore wind; campaigning for 'car free cities' to help local people reduce traffic in their neighbourhoods; promoting the ability to repair things rather than throw them away by supporting two new 'fix-it' factories; developing solar PV schemes that will power railways.[7]

The notion of energy democracy links up with the concept of energy justice, so that the benefits of the low-carbon transition are shared among the poorer people in society. This should include greatly expanding insulation services and offering 100 per cent grants to poorer people to install renewable energy and batteries. Many actions are very cheap to achieve, and financial support can be provided for more expensive measures. Energy democracy also includes incentivising the deployment of green technology developments in poorer areas and those areas which need to transition but are currently dependent on fossil fuels. This is something that is being attempted through the US IRA.[8]

The debate about whether technology or changing behaviour is necessary to save the planet is a false dichotomy. To make sure green energy technologies and systems are deployed, people need to campaign for these changes; this is a change in behaviour. In any case, the discussion seems to implicitly assume that 'technology' is some sort of add-on device to behaviour.

Technology and behaviour are closely related. For example, unless buildings are designed or retrofitted to be energy efficient using heat pumps, it is fruitless just to carry on using fossil fuels as the energy source, no matter how well behaved people are. In transport, unless technology is developed so that, for instance, urban streets favour bicycles over motor vehicles, unless we plan for EVs to take over from ones powered by fossil fuels and unless we plan for electric trams, how people 'behave' is of distinctly sec-ondary importance. Unless our electricity supply is powered by renewables, we shall still be dependent on fossil fuels, and the rate of this change is governed by policy and regulations which need to be reconfigured. These and other changes will only happen if people 'behave' in a campaigning way.

There have been a lot of demonstrations for governments to set more radical targets to deal with the climate crisis and to stop drilling oil. Those sentiments are great, and they set up a context in which demands for technology change are made more powerful. But oil and gas drilling will only stop in all parts of the world when demand for oil and gas dries up. So there need to be a lot more campaigns and actions for practical schemes and policies to ensure green energy technologies and systems are made a priority. If cam-paigning can lead to the adoption of technologies that do not use oil, then most demand for oil will be ended. Then we shall be fighting back effectively in all senses.

CAMPAIGNING FOR DECENTRALISED ENERGY

There are several ways in which energy consumers can fight for green energy, through their own individual economic choices, community initiatives, local municipal initiatives, politics and activism (these things will tend to overlap). Campaigning for decentralised energy is important, involving citizens energy, cooperative and community energy, and municipally organised green energy initiatives. Such a strategy, based on people's choices and actions, is the best way to make progress in the climate crisis.

At its heart, the fight-back for green energy means promoting decentralised energy through bottom-up action. This can take a variety of forms, including:

- setting up community renewable energy companies;
- individuals making their own green technology choices;
- municipal authorities forming and implementing heat plants, to install district heating systems fed by large heat pumps;
- cities supporting the installation of solar PV and batteries in their buildings and infrastructure, promoting renewable energy and making rules that require improvements in the energy efficiency of buildings;
- communities coming together to form their own renewable microgrids;
- the mobilisation of activists towards the adoption of policies that favour green technologies;
- achieving the incorporation of well-organised citizens' assemblies into governmental decision-making processes, which discuss how to achieve green objectives and allow all key interests to make their case to informed randomly chosen citizens – they can then advise governments, rather than merely leaving in place the usual system where corporate interests dominate advice to government; and
- forming campaigns to change the rules and incentives, to make sure green energy technologies are installed.

Campaigners can push for electricity companies, such as cooperative and municipally run companies (in the USA), to become leaders in low-carbon, decentralised energy systems. They are often monopolies in the USA and can be influenced by citizens to promote decentralised energy technologies. One of the most exciting (in terms of potential effects on decarbonisation) but underreported possibilities is action at the municipal level.

PUBLIC OWNERSHIP

Neoliberal and 'modern' conservative approaches have tended to scoff at action by the local state, but for practical purposes the local state is very important, and essential in some areas, for creating and delivering sustainable energy outcomes. Several decades ago, it was accepted by all shades of the political spectrum that local government action was essential for the developing utilities, such as water and energy, that everybody wanted. Having established the basic infrastructure, its administration has all too often been taken away from municipal authorities and given to centralised corporations.

Of course, sometimes control over energy has been left at the municipal level, for instance in many areas of the USA where electricity companies are municipally owned. Yet we are now in a development period, since we must establish the infrastructure necessary for the energy transition. Who is better placed to organise important aspects of such a transition than municipal authorities? This is very much the case when it comes to urban planning and housing policies. We need a transformation of energy demand services, not just changing the mode of supply.

I discussed in previous chapters how transforming the electricity supply to renewables is essential but nowhere near enough, since most of the energy is not currently supplied through electricity. Heat demand, which happens mostly in buildings, is key. Essentially municipal authorities can plan their neighbourhoods to be energy efficient in heating services.

A model for transforming heat supply and demand at a local level was implemented in Denmark in the 1970s and 1980s when local cooperatives were formed to plan, in collaboration with municipal authorities, the creation of district heating systems. They were powered either by existing coal-fired power plants or new (medium-sized) gas-fired combined heat and power plants. These days the fossil fuel energy supplies are being replaced with large-scale heat pumps, so that heat demand can be supplied by energy efficient means. Of course, in most places the cooperative movement is usually not as strong as it is in Denmark. However, the municipal authorities can take the lead, and are doing so in some cases in planning for heat pump-powered district heating systems.

In June 2023, as part of their new heating law, the German coalition government highlighted the importance of municipal authorities in drafting local heating plans. This is to enable low-carbon heating. However, some municipal authorities are ahead of the curve on this, including the cities of Heidelberg and Mannheim. A large-scale heat pump has been installed to serve the existing district heating consumers. The next phase involves connecting other consumers to heat networks fed by large-scale heat pumps.[9]

This has happened because of the intersection of three factors. One is the national consensus for *Energiewende* or the energy transition in Germany. Second is a state-led policy in Baden-Wurttemberg that encourages municipal authorities to prepare plans for supplying sustainable heating. The third factor is the existence of local municipal energy companies, or *Stadtwerke*.

Despite market liberalisation earlier this century, these municipal energy companies still control at least half of energy supplies. This includes the distribution infrastructure, which includes district heating systems.[10] In coming years, it seems likely that there will be an increasing tendency towards expanding the district heating networks and supplying the heat using large-scale heat pumps.

A key factor here is that if municipal authorities run local utilities, then they can respond more directly to local green demands for sustainable energy, compared to the corporate bodies that more often than not control energy production, distribution and supply. German municipalities have been at the forefront of efforts to promote green energy. Some German cities were in the vanguard of promoting solar PV by introducing feed-in tariffs for solar PV as early as 1993.[11] Indeed, this century there has been a wave of examples of 'remunicipalisation' of energy utilities in Germany, perhaps the biggest being the city of Hamburg.[12]

In the UK, where local authorities ceased to own energy infrastructure after nationalisation in 1948 (and privatisation after 1990), some district heating systems have been organised in collaboration with private companies. It is still possible for local councils to organise district heating systems, but it is an occasional activity compared to the development of district heating in

Germany. The fact that in Germany 14 per cent of households are linked to district heating companies, compared to only 2 per cent in the UK, may have something to do with differences in municipal control of energy.

In the UK there have been attempts to set up municipally owned energy suppliers, but these have just involved supply rather than distribution and (much) generation. Moreover, British local councils have to apply to central government for small pots of money made available for capital investment in district heating systems. Currently there is nothing available in the UK for large-scale heat pumps to serve district heating systems.

Municipal authorities should be given mandates to establish plans for heat networks served by large scale heat pumps. Municipalities should have access to sufficient funds to pay for such plans. This is a crucial way of saving massive amounts of energy and helping consumers, especially in areas where heat pumps in individual houses are less practical.

There are many municipal energy authorities in the USA but, as mentioned in Chapter 3, they have (until the very recent IRA) been prevented from accessing federal tax incentives for renewable energy. This appalling bias towards the big corporations has held back the prospects for an energy transition in the USA. Municipal authorities have been able to buy in renewable energy by signing PPAs with renewable energy generators. However, they will be paying the full market price for the energy, while the private generator keeps the value of the PTCs that will have been used to fund the renewable energy investment. However, with the IRA they will be able to receive 'elective payments', meaning that they can obtain cash support (to invest directly in green energy) equivalent to the tax credit support to which they were previously barred.

In a press release, the American Public Power Association (APPA), which represents municipal electricity utilities, said:

Prior to IRA, utilities serving nearly 30 percent of the nation's customers were excluded from receiving energy incentives delivered through the tax code, meaning the vast majority of wind, solar, and other non-hydropower renewable generation is owned by merchant generators with roughly 60 percent of the

value of associated tax credits going to banks, insurance companies, and other financial 'owners'. ... Elective payment of tax
credits has the potential to be revolutionary for such investments, unlocking the ability for public power communities to
own and control such projects, rather than going hat in hand to
Wall Street hoping to find a willing investor. That means local
decision making driving local generation and jobs.[13]

Hopefully now we shall see the forces of 'public power' unleashed
to rapidly increase the quantity of green energy projects in the
pipeline. Unlike in Germany, there have only been a small number
of remunicipalisations in the twenty-first century. However, with
the change in access to funds to promote renewable energy, more
could be available if municipal companies could offer a more rapid
path to decarbonisation through developing green technologies.

In fact, some publicly owned energy utilities did try and
advance the green energy agenda as much as they could within
these discriminatory financial constraints. For example, Gainsville municipal electricity utility (in Florida) offered a feed-in
tariff for renewables, including solar PV. The Tennessee Valley
Authority also offered feed-in tariffs. This is a lot more than the
'investor-owned' utilities managed, and this reflects the ability of
public authorities to be influenced by the desire of voters to favour
renewable energy.

An important way in which publicly controlled energy utilities could help build a system based on renewable energy is by
ensuring that the lower costs of renewable energy contracts are
passed through to consumers. City authorities themselves can act
in various ways to shift corporate power and to shape outcomes in
the direction of a green energy transition. Different cities in different states have different powers over building codes, for example.
In the USA and Canada, in some states/provinces cities have power
over building codes. Councillors in Vancouver, for example, have
attempted to change city policy to ban gas connections for new
buildings.[14] This initiative did not fully succeed, but such actions
add to pressure at a provincial level.

Several cities in the USA have pledged to move towards a 100
per cent renewable energy supply. Such efforts can be valuable

if they help schemes start up that would not otherwise be developed. A warning here is that sometimes these can be good-looking but meaningless actions, if all that is happening is that contracts are being signed with existing renewable energy schemes, or with schemes that would be happening anyway, without the city's nominal support.

ROOFTOP SOLAR AND BATTERIES

Building as many large-scale (utility-scale) solar farms as quickly as possible is important, but rooftop solar can make, and in some countries is making, a major impact. Putting solar PV on buildings is a popular activity when there is a strong groundswell of people's enthusiasm for implementing rooftop solar. Indeed, the countries with the fastest-expanding solar power programmes lean heavily on rooftop solar.

This includes Australia and the Netherlands, both of whom derive the largest share of solar generation from rooftop solar. These countries lead other large, industrialised nations in terms of generating the most solar PV per person. Decentralised energy is a problem for energy corporations since it is not owned by them and takes away their business. In the USA, consumers face mountains of red tape if they want to install solar PV, and this increases costs greatly for people who want to install solar PV.

In Australia, rooftop solar is overhauling coal as a source of electricity. Solar PV generated nearly 14 per cent of electricity in 2021, nearly two-thirds of this coming from rooftop solar. These countries are well ahead of countries such as the USA and the UK in terms of rooftop solar installations. Perhaps the reasons for this can be expressed through three factors: planning and connection rules, incentives and culture.

Of course, these three factors interact with each other. If you have a culture that sees solar development as important, then planning and connection rules are likely to be permissive for solar, and incentives are likely to be given out by the state. This is the case in both the Netherlands and Australia. There are no complicated rules for getting consent to putting solar panels on roofs. As discussed in Chapter 3, such rules are a killer in the USA, by dramatically increasing costs and slowing things down generally. In the Netherlands there are generous net metering incentives, and in Australia local states give incentives for solar development. The Netherlands has a strong cultural drive to do things that counteract climate change. The Netherlands is a low-lying country that is very vulnerable to increases in sea levels. In Australia, there is a culture of self-sufficiency that an article in the *New York Times* calls 'rugged individualism'.[15]

As discussed in Chapter 3, in many parts of the USA the cultural pressure in favour of renewable energy has not yet been able to overcome the systemic institutional bias against solar PV which weighs down installations with expensive, time-consuming and irrelevant safety checks and planning applications. What has been possible is that policies at a federal level have now (through the IRA) produced tax incentives to be extended to residential solar PV. In addition, there has been the longstanding institution (since the 1970s) of the federal PURPA laws, which has meant that corporations have been forced to offer some level of payment for solar exports into the electricity system.

However, there is a part of the USA that has been able to reduce the planning and regulatory hurdles to installing solar PV, and that is Hawaii. This effort has focused especially on inducing the corporate utilities to ensure a streamlining of the process of deploying solar PV. In contrast to the USA's general lack of cultural pressure to overcome the obstacles to solar PV, in Hawaii there is the pressure to rescue the state from its high energy prices. It is very expensive to ship conventional fuels to Hawaii, and as a result the state has set installation of solar PV as an important policy to secure its vital energy security needs.

Hawaii is mostly reliant on expensively shipped oil products to provide energy (and electricity largely from diesel generators), although consumption of renewable energy is now increasing. Residents in Hawaii have tended to face very high energy prices compared to other places anyway, and since 2021 these prices have become worse and worse. Hence there was great pressure on the government to help consumers. Supporting solar PV was an obvious way of doing this.

Among the support given to solar PV are large grants to enable homeowners to install batteries – something that very much helps the energy security of the islands. 'Nearly a third of single-family homes' have solar panels fitted.[16] The time-wasting and expensive bureaucracy that faces solar installers in other parts of the USA is being torn away after the Hawaiian Public Utilities Commission (HPUC) issued 'performance-based regulation' to ensure the barriers to solar installations – and also energy efficiency – were removed. The HPUC said:

> The suite of new performance mechanisms includes an Interconnection Approval performance incentive mechanism ('PIM'), which incentivizes faster interconnection timelines for small-scale solar and storage systems, and an LMI [low-to-moderate income] Energy Efficiency PIM, which incentivizes increased collaboration between the utility and the energy efficiency programme administrator to provide low-to-moderate income customers with opportunities to better manage their energy consumption.[17]

So why can't this approach be replicated across the USA? Solar advocates such as the Southern Alliance for Clean Energy are pushing for regulatory policies that include giving premium export prices for residential solar units that also have home batteries. This is win-win situation for efforts to integrate fluctuating renewables in the electricity system.

COMMUNITY RENEWABLES

As discussed in earlier chapters (especially Chapter 1 and Chapter 4 on the EU), modern renewable energy started largely from community renewable initiatives of one form or another. In Denmark, the emphasis on renewable energy is based on widespread community and public consultation and a tradition of cooperative energy initiatives, including cooperative ownership of wind farms. In Germany, much of the renewable energy capacity is owned by ordinary people.

However, since then big energy players have taken over in deploying most of the renewable energy. Given the need to achieve greater public acceptance, and also extra renewable energy capacity, it is important that the number of community renewable energy projects is increased.

Community renewable energy comes in various forms. It could mean ownership by local people in general, which could mean local farmers. That is great. However, here my focus is on other forms. Community renewables can mean cooperatives, whereby community-oriented groups or companies develop projects themselves, which are owned by many small shareholders. They will earn revenue from the profits from income that is earned from renewable energy generation. It can also mean that specialised companies sell shares in renewable energy projects, or it could mean that conventional, large-scale renewable energy projects are part owned by ordinary subscribers. This latter type will result from a developer selling a portion of shares to the public. Community ownership of renewable energy has several social benefits. It turns more people into active supporters of renewable energy and renewable energy projects since they have a stake in the technology. It often also results in community projects helping local causes such as relieving fuel poverty and supporting energy efficiency initiatives.

COMMUNITY RENEWABLES IN EUROPE

In fact there is a growing tendency, in Europe, for legislation to stipulate that in some circumstances, shares in renewable energy projects must at least be made available for purchase by members of the general public. In the USA there is a growing amount of interest in 'community solar' schemes. In these cases, opportunities are given to people who cannot easily own solar power themselves to buy shares in community solar parks. Also, there are many examples of pure community renewable energy projects (increasingly solar) which are funded by selling shares to individual enthusiasts. Many of the community renewable projects fund local causes including energy efficiency and anti-fuel poverty measures.

There has been a strong European history of cooperative ownership of renewable energy. As discussed in Chapter 4, the earliest wind power programmes, from the 1970s to the early 2000s in Denmark and Germany, were dominated by citizen ownership of wind power of one form or another. This could mean, as in Denmark, pure cooperatives where all of the shares in the company that owned the project are owned by local people, or farmer cooperatives where local farmers ran the cooperatives or owned the wind turbines themselves. Alternatively, in Germany there have been these types of cooperatives or (covering the bulk of wind deployment in the past) hybrid companies. In this case, small corporate bodies derived their investment by selling shares to the general public. However, in the age of big offshore wind farms, the notion of citizen ownership lapsed. It is now being revived in a different form.

The EU has set out some plans for encouraging community ownership. This includes putting more emphasis on the original design of feed-in tariffs to support community renewable projects. Over the last decade, EU policies have favoured the use of auctions for fixed price renewable energy contracts. This means that companies compete, in terms of bidding prices for what they would be paid per MWh generated for the period of a long-term contract or PPA. The winners are the companies who bid the lowest prices for a given amount of volume of capacity offered by government agencies in a particular bidding round (typically annual or bi-annual).

The auction system produces cheaper prices paid to big corporations and this is good for the consumer, in that it keeps their bills down. However, the system discriminates against community energy schemes, in that they find the administrative requirements (including the requirement to trade on wholesale power markets) prohibitively expensive. Traditional feed-in tariffs involve a set price to be paid for renewable production by the government or an agency charged with doing this.

However, the EU initiative goes further than just supporting the reinstatement of feed-in tariffs for small projects. Community and cooperative energy schemes have political, economic and social advantages. In economic terms, they can source investment from ordinary people, and thus put more money into renewable energy businesses. Second, they can be cited as renewable energy schemes that involve the grassroots and not just big companies. Third, they can be sources of innovation. New techniques can be tried out by community-based and -funded schemes that might not be financed by conventional investment streams. Fourth, they can spread the economic benefits of the renewable energy investments around.

A leading example of community participation, which follows from the EU initiative supporting community energy, is the Renewable Energy Community initiative organised by the Belgian government for its latest offshore wind schemes. As one of the governmental consultation documents says:

> The purpose of the integration of renewable energy communities is to ensure that citizens do not only participate financially in the offshore wind projects, but that they can also get a high level of active involvement in the development, operation and decision making process of the project and access to the offshore wind electricity ... This would allow citizens to benefit directly from the low cost price of renewable electricity. The ownership structure of renewable energy communities can help in increasing the uptake of renewable energy installations where individual investments are less likely, leading to democratization of access to renewable energy resources, while also contributing to the reduction of energy poverty.[18]

This particular tender involves up to 3.5 GW of offshore wind power. The citizens' investment will make up 20 per cent of the shares in the offshore wind farm. In addition, the wind farm will supply 20 per cent of the energy it generates to individual Belgians.

This initiative arose in the wake of campaigning by a collection of renewable energy citizens' cooperatives, who have constituted an organisation called Seacoop.[19] In this example we can see a virtuous interaction of pressure for citizen-based cooperation at the interstate level (EU), national level (Belgian government) and grassroots mobilisations through the dozens of renewable energy cooperatives who formed Seacoop.

In the Netherlands, this principle of citizen ownership of shares in renewable energy is also being extended to solar farms. There, shares are offered to local communities, and there are also crowd-funding opportunities whereby people can invest in solar power projects in places other than where they live. In fact, such opportunities are available, to a greater or lesser extent, in various countries. In the UK, quite a few community solar schemes were set up in the 2010 to 2019 period when the UK's feed-in tariff for small renewable energy schemes was still operating.

Since then, it has become much more difficult to organise community solar schemes since there is no incentive scheme available for smaller projects. The government does award some long-term PPAs for schemes through competitive auctions, but community solar schemes are not offered the level of rates won by these schemes. Clearly there could be a community renewables programme, at least for small projects, and they could be organised to engage in innovative activities. Such innovation could include incorporating 'agrivoltaic' schemes, wherein agriculture is combined with solar farms.

However, some people are trying to organise community renewable energy schemes even in the present circumstances. For example, an organisation in which I am an investor, called the Big Solar Co-op. It is a business that is funded by shareholders in the community that installs solar PV arrays on commercial premises. The commercial premises effectively rent out their roofs to allow solar panels to be installed. These generate electricity which is sold back to the owner of the building at lower rates than what they would pay to the conventional electricity supplier. The roof owner has the option of buying back the solar array at what Big Solar Co-op describes as 'cost' terms. Big Solar Co-op also has the advantage of an enthusiastic volunteer group. They can do various things, from recruit and help organise sites for solar arrays, to even designing the arrays. I have pursued a few leads myself, although until now these have not led very far. But I'll keep trying! The company itself is generally doing well, and is moving ahead with several projects.

A more direct community renewables route in the UK is operated by Ripple. They own or part-own wind turbines and solar farms, and subscribers, in turn, buy shares in the initiative through Ripple. The scheme works by Ripple having agreements with electricity suppliers, whereby investing consumers receive credits for their electricity bills according to what their 'share' in the renewable energy project has generated. In fact, this type of initiative is close to the 'community solar' concept being implemented in the USA.

COMMUNITY SOLAR IN THE USA

Community renewables in the USA are mostly focused around community solar. This is organised in different ways, but it is oriented, in particular, to allowing people who do not have the possibility of putting solar on their rooftops to effectively own solar panels somewhere else. At its most simple, this can involve having a solar array connected to the apartment building where somebody lives. People can buy shares and are remunerated (in proportion to their shareholding) with the income that is gained by selling the electricity back to the residents. Then there are (usually larger) schemes organised by special purpose companies that sell their shares to people in the community who receive a share of the profits generated by the scheme selling the electricity back to a utility.

However, the most prominent type of community solar scheme is organised through state regulatory authorities. These allow subscribers to the scheme to be rewarded through the 'net metering' process. This means that people can earn income from the scheme (in which they have invested) by being paid as if the portion of the generation that they own was in fact on their roof (if they have one). A variation on this theme is that rather than investing a lump sum at the start, people can pay a monthly fee which will be smaller than the returns they will receive because of the solar generation.

So far, by early 2023, around 5.8 GW of community solar has been deployed in the USA, with around 11.5 GW expected by 2027. Community solar projects have been established (by 2023) to a greater or lesser extent in 41 US states.[20] There are 20 states which have established formal mandates that drive utilities to offer community solar schemes.[21] The IRA will make setting up community solar projects easier (as it does other publicly owned renewable energy initiatives) since, in effect, the ITCs for renewable energy can be turned into cash. There is a big effort, led by organisations such as Vote Solar, to build up commitments in US states to develop community solar schemes. The projects themselves no longer have to have a tax liability themselves. However, big limitations on the speed of setting up solar schemes are the grid constraints that face renewable energy projects in general in the USA.

MICROGRIDS AND PEER-TO-PEER TRADING

One way of fighting back against corporate control and fossil fuel domination is to organise 'microgrids' and 'peer-to-peer' trading. These are different things in technical terms, but they share the common feature of avoiding the conventional pathway of power stations supplying energy to passive consumers. Community groups are themselves directly empowered by the notion of peer-to-peer trading whereby community renewables projects can sell directly to consumers. In the UK, an enterprising pressure group called Power for People has been making progress in promoting legislation in the UK Parliament to allow for this. It will mean that local people organising, for example, community rooftop solar projects should be able to sell their power at good rates, rather than the often low rates they will be paid by electricity suppliers.

A rather more technically complex form of energy decentralisation is the notion of the microgrid. This is a form of 'virtual power plant', where a locality can supply as much of its energy needs as possible from local energy assets. These will comprise local renewable generation, batteries and other types of energy storage, and various demand-shifting techniques. The whole thing is operated automatically through digital controls.

As EVs become more prevalent, so-called bi-directional charging will become ever more important. That is because EVs have large batteries that not only soak up power from renewables when it is in surplus but also the newest EVs on the market can send power back into the building. This means that in sunnier places, in theory, communities should be able to be self-sufficient from any external grid, so long as they have enough locally installed renewable energy resources. Together all of these assets are known by the technical term 'decentralised energy resources' (DER).

Although they are generally promoted by hi-tech companies, in fact microgrids can be organised by even the poorest communities (given access to some capital). In Puerto Rico, following the wipeout of energy systems in Hurricane Maria in 2017, locals in the town of Adjuntas took action and formed the Community Solar Association of Adjuntas. They were building on previous solar initiatives, but they have established a small microgrid connecting various buildings in the town centre, with batteries and solar arrays. This was able to keep everybody's electricity services going despite Hurricane Fiona in 2022.[22]

In fact, there are strong arguments to say that with extreme weather events increasing because of climate change, DERs can greatly improve grid resilience. According to an analysis published in *PVMagazine*:

> [A recent report] showed virtual power plants, such as local solar and storage, can provide the grid with 40–60% lower cost resource adequacy. These new technologies can improve the grid's bottom line and reverse the trend where consumers are on the hook for decreasing grid reliability and higher electric bills. ... Not only that, but DERs like community solar projects, neighborhood microgrids, virtual power plants, and income-based energy efficiency programs can lower the energy burdens on low-income communities and deliver on environmental justice. ... Solar energy and battery storage technologies can revamp the grid into something that is cheaper, more stable and more equitable.[23]

Unfortunately, *PVMagazine* also had to report that efforts to develop microgrids in new buildings are being rejected by the regulatory authorities in California. The is after pressure from the mainly monopolistic utilities themselves. Paradoxically, the position adopted by the California Public Utility Commission is that they are preventing a 'monopoly' being established. This is ironic since a microgrid can protect residents from the effects of corporate monopolies. According to *PVMagazine*, the company that wants to develop the microgrids, Sunnova, 'seeks to develop largely self-sustaining micro utilities by equipping new home communities with solar and storage. Under the plan, new home construction communities are selected so that Sunnova can work closely with developers to design and implement distributed solar microgrids backed with energy storage.'[24]

CITIZENS' ASSEMBLIES AND DELIBERATIVE POLLING

A powerful tool is to organise and have institutionalised citizens' deliberation for energy and environmental issues. The examples discussed here are from the USA, France, the UK and Ireland.

I stress here that I am talking about citizens' discussions that are incorporated into an official decision-making process and which are designed so that all relevant interest groups have an input. I do not wish to dismiss citizens' assemblies' organised by activists – they are good as a focus for discussion, but to have more credibility as a policy-influencing device, they should be organised on a rigorous basis. Of course, even then an antagonistic politician can complain that they 'don't want to be told what to do' by a citizens' assembly. However, if they say this, they are being disingenuous since the consultations still leave it up to the politician what to decide. It just means that the politicians do not get told what to do by the corporations (on their own), as is usual practice.

The advice from institutionalised citizens' deliberation discussed here is advice that is more truly in the public interest compared to the current, corrupt system of law-making. Elected politicians are still left to make the final choices, but the sole influences on such choices are no longer the corporations who want to preserve their own power and profits. The choices, made by informed citizens, are more likely to work to achieve climate change objectives than policies adopted mainly to suit the demands of powerful interest groups.

These various mechanisms of citizens' democracy and control (which constitute energy democracy) are powerful in illustrating how ordinary people, in the context of being informed by the interest groups and experts, can give policy advice and take direct control over energy themselves.

According to Rebecca Willis Nicole Curato and Graham Smith:

Simply put, deliberative democracy places reasoned discussion at the heart of democracy. ... Deliberative democrats ... challenge the view that citizens are too ignorant, or too disinterested in politics. ... In terms of decision-making processes, deliberative democrats stress the importance of considered judgment, based on good evidence and free and fair collective discussion.[25]

TEXAS

The technique can be, and has been, deployed to gauge public attitudes on particular issues, including this example of renewable energy in Texas.

A deliberative poll in Texas spurred on efforts to set up the renewable portfolio standard. The poll was initially established after power companies in Texas asked the political scientist (and advocate of deliberative polling) James Fishkin to report on citizen attitudes to renewable energy.

Hundreds of people were selected randomly and contacted via telephone. They were initially polled to see what their views on the issues under discussion were. They were then asked to attend a weekend of deliberations. They were given information regarding what was designed in a non-partisan manner about the policy and technological options. Then the attendees were given a second poll about the issues.[26] According to Ted Wachtel (a climate activist):

Texans – from the gas and oil state, who drive more miles in more pickup trucks and SUVs than folks in any other state – were willing to pay extra money for renewable energy and for energy conservation. From the telephone poll to the final poll, customer willingness to pay extra money jumped 30 percent, to 84 percent for renewable energy and 73 percent for energy conservation.[27]

These results, which surprised people, were publicised in the media. They had political influence, and a result of the policy discussions was a relatively early regulatory boost for renewable energy in Texas.

CLIMATE ASSEMBLY UK (2020)

As a result of a decision by six committees of the UK Parliament, a Climate Assembly UK (CAUK) deliberation was established and organised by the official Committee on Climate Change. 'Climate Assembly UK brought together 108 people from across the UK and from all walks of life to examine the question: How should the UK meet its target of net zero greenhouse gas emissions by 2050?'[28]

The representative sample of people were informed by a series of briefings coming from various points of view about the main options for greenhouse gas reductions. They then had discussions and made representations. The whole process was overseen by the Committee on Climate Change, in collaboration with a number of senior academics.

They looked at the briefings to ensure they were balanced. CAUK made a series of assessments based on the votes taken by its members. Then the assembly made a number of detailed recommendations, which were published.

CITIZENS CLIMATE CONVENTION IN FRANCE (2019–20)

Arguably the French climate deliberation, named 'Convention Citoyenne pour le Climat' (CCC), was not only better empowered than CAUK but was also much more politically engaged and relevant. As noted by Catherine Cherry and her co-authors:

> Whereas CAUK participants were supported to be independent and as representative of 'ordinary' people as possible, CCC participants were actively encouraged to engage with politics. It was not suggested to CAUK participants that they undertake additional research on the topic of climate change in between sessions, or to have wider conversations with friends and family about the issues at hand.[29]

Cherry et al. also noted that the CCC received much higher levels of publicity than its British equivalent and its output was more directly plugged into the political process.[30] Whereas the CAUK was established purely as an advisory body, the French assembly was set up directly through the offices of President Macron. It made proposals directly to government, some of which were accepted and some of which were rejected.

CLIMATE ASSEMBLY IN IRELAND

Ireland has been an early mover in citizens' assemblies. The Irish system involves a call for submissions from the public, which are considered by a randomly selected body of citizens who deliberate over two weekends. They are supplied with expert policy briefs. They then vote on a series of recommendations. The results of the deliberations are considered by a Parliamentary Commission.

It has organised them in a series of policy debates, starting most famously with a deliberation on reforming abortion law. This led directly to a referendum on the subject. However, climate change is also regarded as being one of the more effective topics for a citizens' assembly. According to Coleman et al., 'the parliamentary committee's report shaped to a significant degree Ireland's landmark Climate Action Plan'.[31]

DANISH CLIMATE CONSULTATION (2008–2010)

The Danish commitment to move towards a fossil-free (and energy independent) energy economy by 2050 began in 2008 with the appointment of a Climate Commission, and then opened out into a public consultation. The process ended with legislation in 2010.

The Climate Commission consisted of ten academic scientists covering relevant areas which examined how to phase out fossil fuel use in Denmark. According to a paper by David Toke and Helle Nielson:

In the fall of 2010 the Commission produced 40 recommendations, based on a large body of scientific background analyses and modelling exercises. The Commission concluded that the objectives of a fossil-free energy system and significant CO_2 reductions could be achieved at a relatively low cost and primarily through the promotion of renewable energy and increased energy efficiency.[32]

The point about this exercise is that it preceded the formation of government policy and legislation and was led by a panel focused on the climate issue. Although this was a more top-down arrangement compared to the three earlier examples of public deliberation, it was driven by a focus on the best means of achieving climate and energy independence objectives, rather than what the corporations think is in their interest.

It was done independently of the big energy corporations who usually dominate energy policy formation. Indeed, in the years following this exercise, the publicly owned DONG changed into a purely renewable energy company: Orsted.

GOVERNING FOR PEOPLE BEFORE CORPORATIONS

There is a possibility that once the corporations are weaned off fossil fuels in favour of renewable energy, they may try to charge the consumer higher prices for renewable energy generation compared to what it costs them to buy it. I have already discussed some means of avoiding this outcome. The corporations are fond of saying how things would be better if they are left to the market – which means markets that they control – rather than for the benefit of the people. In Europe there are auctions for contracts for renewable energy that mean that developers are paid a set price for the project's generation. This is a good way of allocating contracts for renewable energy. Governments can plan for the amount of renewable energy that is needed, and the price that is paid for the renewable energy is enough for the developer, without the companies that own the project being paid excessive profits.

But corporate players appear not to like this, because they sense that they could earn more money if prices were left up to the market. In this way, neoliberal language is used to persuade people into thinking that things will work out more efficiently if governments do not set precise prices. In fact, this is not the case. When energy prices are high, as they were in 2022–3, the companies can sell the renewable energy for higher prices and take bigger profits.

This became evident when some offshore wind developers, with projects that had not yet been started, commissioned the projects without taking out the British government's CfDs. These CfDs stipulated much lower prices for energy generation than the then bloated market price. The British government has since made proposals to close the loophole that allowed them to do this. In future, the developers will have to accept the CfDs and their level of payments when they start the projects.

The moral of this story is that governments and regulators should issue fixed price contracts for renewable energy, that is contracts which pay fixed prices for energy generation. In doing so, governments can plan for how much renewable energy needs to be developed to meet climate and energy security targets. On top of this, governments need to plan and develop balancing and storage facilities to run the new renewable energy supply system. Finally, governments should own the retail energy suppliers to ensure that the savings from cheap renewable energy are passed through to the consumer.

We need corporations, both public and privately owned, to organise big onshore/offshore wind and solar PV projects as well as big efforts to promote community-owned renewable schemes. This should happen with contracts issued by governments which deliver cheap energy for the consumer rather than excessive profits for the corporate shareholders. The UK government developed a good system for issuing these CfDs through auctions for the contracts. The only problems arise when not enough are issued and when community renewable energy companies are not offered access to contracts.

Yet in the battle between the big energy corporations and the people, the signs are not good in the case of German offshore wind energy. In July 2023, large amounts of offshore wind capacity were auctioned off by the German government, with the winners of the auctions (oil companies as it happens) making negative bids for the projects.[33] Superficially, this looks marvellous in the sense that the consumer is apparently promised a lot of money back. But Jerome Guillet, a very knowledgeable independent offshore wind consultant, has criticised this deal. Guillet says that the companies may only develop the projects if either the market prices are very high or there are (additional) public subsidies thrown to the developers. He comments:

> For oil companies, the bets on offshore wind, however large, are still massively dwarfed by their investments in the traditional oil & gas sector, so they don't really care about the eventual monetary losses (net of the public subsidies they manage to scrounge) and they are worth it just for the PR value, and maybe the option value. ... For utilities or renewables pure players, it's a tougher situation – be disciplined and lose out on generation capacity (and shrink), or bet against the oil companies, with the risk of losses that are much more visible ... the benefits of the largely fixed cost of production will not go to the general public, contrary to what would happen if you had a CfD.[34]

State-owned renewable energy generation corporations should be created to increase competition in renewable energy development. This has been proposed by the British Labour Party. This currently exists in the shape of Orsted, in which the Danish government owns a majority share.

On the other hand, public ownership of the retail supply of energy will itself bring benefits for two reasons. First, this can ensure that the savings gained through well-priced renewable energy PPAs are kept by the consumers themselves rather than the corporations. Second, publicly owned suppliers have the potential to organise DSM better than suppliers in liberalised retail supply markets. Consumers need to be paid to shift their demand from one half-hour to another and to invest in equipment such as batteries which helps this to happen. A publicly owned retail electricity supplier can share around these costs. Competing retail suppliers will not wish to incur extra costs that may benefit the national or local energy distribution system rather than their judgement of their own competitive position.

Public ownership of retail supply can also ensure that the savings gained through well-priced renewable energy PPAs are kept by the consumers themselves, rather than kept by the corporations.

Last, but most importantly, the process of electrification of the economy (and shifting away from oil and gas) needs to be promoted through well-organised regulations and incentives. These need to encourage heat pumps to be installed as rapidly as possible in buildings with much improved energy efficiency. Regulations and incentives need to ensure that fossil fuel-based transport systems are phased out as quickly as possible.

CONCLUSION

Energy corporations will pursue energy transition objectives only in so far as it is in line with their core interests. Existing fossil fuel interests, and interests in general in centralised solutions such as nuclear power, act as barriers to a rapid energy transition that maximises the benefits for consumers. Big companies are needed to organise major renewable energy projects, but CfD contracts need to be used to maximise consumer benefit. Public ownership of suppliers can help to ensure these consumer benefits materialise.

There are a number of ways that corporate power can be challenged, or at least channelled into consumer-friendly paths to sustainable energy transition. Essentially this can happen through two means: greater government intervention and energy democracy. Energy democracy has been defined in various ways,[35] but here I mean that ownership and control of energy is diversified and decentralised, and that ordinary people are included in the policy decision-making process. An immediate way for people to own and control energy assets is through being 'prosumers' – both producing and consuming and storing their own energy. Allied to this are the efforts of campaigning organisations focusing on pushing for policies to support the green transition: technology social movements. Altogether these constitute the social movements that push forwards the energy revolution.

The simplest and most direct means of energy democracy is home ownership of energy assets – including solar PV and batteries. Community energy schemes are a continuing means through which citizens can increase their own influence and democracy in energy ownership. These have been backed by legislative means in various countries. In the USA, community solar schemes now have access to tax credit support, but in other countries, such as the UK, community renewables projects do not have access to the financial support enjoyed by corporations who develop renewable energy projects.

Community renewable projects can also empower those people who are not able to put solar PV on their roofs to have a share in owning renewable energy. However, without the legislative support (such as that provided in some states in the USA for example) those citizens may not be able to do this (especially in the case of the UK). In addition to this there are, and should be, efforts to ensure that a portion of the shares is offered to the public, in what are otherwise corporately organised renewable energy schemes.

Greater democratic control of the policy process itself – and more effective and more rapid transition – can be achieved by deliberative, democratic means. Such systems can ensure that the corporations do not have a monopoly over political decision making. There are different designs that might achieve this. However, among the examples presented, a key facet is that the policy discussions, whether led by independent panels of experts (as in the Danish example) or by independent panels of citizens (in the French, British and Texan examples), need to be well informed by all relevant interests. They need to have space for discussion, debate and consultation, and the results of their deliberations need to be well transmitted into the political system and relevant (energy) debates about policy and regulations.

There is a separate dimension whereby public institutions, for example municipal energy companies, can play an important role in activities such as developing heat networks powered by large-scale heat pumps. Locally owned energy companies are ideally suited for this purpose since they already have housing and building respon-sibilities. Municipally owned energy companies that already exist can be given greater powers to invest in renewable energy. It is very anomalous that in the USA, municipally owned electricity compa-nies have been denied the incentives open to corporations to invest in renewable energy.

Other forms of public ownership can also benefit the energy transition. The creation of publicly owned companies, designed to develop renewable energy projects, can increase competition with the existing multinational corporations. Indeed, some of the energy corporations in northern Europe are themselves publicly owned.

The public ownership of assets in retail energy supply, where there is no meaningful competition, can also benefit consumers and the energy transition. A publicly owned retail supply can assure that all consumers are offered favourable terms to shift their demand in order to balance supply from fluctuating renewable energy sources. A combination of publicly owned retail supply and use of the CfD system for awarding generation contracts to renewable energy developers can ensure that consumers are not charged extra (as pure corporate profit) for renewable energy over and above what it costs to buy the renewable energy from the renewable generators.

6

Conclusion

100 PER CENT RENEWABLES: SOONER OR LATER?

The current trends in rates of growth of renewable energy point towards 100 per cent of the world's energy being provided through renewables by 2050. That would achieve the terms of the Paris Agreement for global net zero by 2050. The green energy revolution will be cheaper than opponents of green policies suggest. Wind, solar and electric cars, for example, are becoming ever cheaper – much cheaper than fossil fuels or nuclear power.

However, these objectives depend on policy breakthroughs being made. Currently, even very cheap measures, such as banning fossil fuel equipment in new buildings, promoting EVs, making solar PV mandatory on new buildings or giving good contracts to renewable energy projects are being either delayed or successfully blocked in many places. Incentives need to be greatly increased and infrastructure spending boosted to spur on the green industrial revolution. Finance for green energy needs to be given in large quantities to the poorest countries in the world.

Besides renewables, there needs to be a major shift towards electrified, energy efficiency technologies (including heat pumps and electric transportation), planning for a big expansion in grid capacity and different types of short- and long-term energy storage. This requires vigorous government intervention, including public ownership to deliver it where appropriate. We need to ensure that the profiteering by energy corporations from fossil fuels is not transformed into profiteering from renewable energy as the energy transition progresses.

In Europe neoliberalism has failed to protect energy security, and indeed worsened the impact of the gas crisis compared to less market-led energy arrangements. Moves towards the energy transition have only been possible by making green energy technologies effectively exempt from liberalised energy markets. This has been done by offering guaranteed long-term prices for renewable energy generation and issuing regulations and incentives to support electrification and energy efficiency. Meanwhile, in the USA the power of energy corporations and their interests in fossil fuels has often prevented renewable energy projects being given priority. On the other hand, struggles by individual North American cities and some states for green energy are providing some hope of introducing energy efficient technologies and growing renewable energy.

THE FOSSIL ENERGY CORPORATIONS:
A DINOSAUR PROBLEM

The problem with the energy corporations is that they often carry on doing what they are used to doing, despite the need for a paradigm shift in energy production and use. Meanwhile the oil and gas corporations have amassed stupendous profits, only a small fraction of which is likely to be invested in green energy technologies. The corporations will service their current assets (and existing, now dated, expertise) as much as possible following their existing business model.

With some notable (mainly European) exceptions, the energy corporations have often seen renewable energy and decentralised energy as threats to be warded off, or at least stalled. It is only more recently that the electricity corporations have turned to replacing their older fossil fuel power plants with increasingly cheap solar and wind power. The earlier development of renewable energy has been left first to enthusiasts and then to independent companies.

In the long run, the (green) technological solutions that work are driven first by people and environmentalists and then optimised by industry. Despite the obstruction by energy corporations, great strides are now being made in deploying renewable energy and increasing proportions of electricity are being supplied by them. Heat pumps, much better insulation of buildings, EVs and other forms of electrified or people-centred transport (such as bicycles) need to be promoted much more vigorously and planned by government at all levels. The big energy corporations will help so long as they are made to do so by governmental intervention.

Corporations try to promote solutions such as 'blue hydrogen', CCS of fossil fuels or even nuclear power. These solutions divert money and political attention towards technology dead ends. Reliance on dubious systems involving fossil fuels are being pencilled in as a key means to balance fluctuating renewable energy rather than using more sustainable means of balancing fluctuating renewables. Worse still, and this is a feature of US energy politics in particular, emphasis is being put on investment in new gas-fired power plants without even carbon dioxide removal. The only realistic method is to base strategies on achieving a 100 per cent renewable energy supply.

NUCLEAR DIVERSIONS: OUTDATED DISTRACTIONS

In general, energy corporations have a declining interest in nuclear power, for commercial reasons. Nuclear power is associated with what is now, in the West at least, a bygone era of heavy industrial construction projects which were made feasible in the past by a large cadre of heavy engineering labour. This involved both specialists in areas such as welding and a large pool of cheap labour, neither of which exists in the West any more, or at least not in large enough quantities. This heavy engineering labour force used to work in the context of far less stringent, and therefore cheaper, health and safety regimes. New nuclear power is designed for greater safety which also increases costs.

While renewable energy technologies improve their efficiencies through design optimisation and supply chain rationalisation, nuclear power is locked into twentieth-century construction techniques. It is further limited by the (low) efficiencies of steam-powered turbines. These turbines waste the bulk of their energy (though it is still counted as nuclear production by the IEA). Wishful thinking is being promoted about fantasy SMRs based on technology that either exists (as in nuclear submarines) that is far too expensive for civil power production or was abandoned around 60 years ago because it was impractical.

Further evidence that nuclear power should belong to the past is the fact that Russia puts so much store in it. Nuclear power's geopolitical associations are dubious given that Russia is the pre-eminent exporter of the technology. Russia influences governments through the political lock-in associated with its nuclear export schemes. Rosatom (Russia's composite civil and nuclear company) ties donor countries to dependence on Russia for future operation and repayment of loans needed to build nuclear plants. Should we encourage proliferation of nuclear power if there is a plausible and arguably just as cheap (probably much cheaper) pathway that does not involve nuclear power? My answer would certainly be no. However, regardless of my views, nuclear at best seems capable of providing no more than a trivial proportion of the world's future energy supply. The analysis in Chapter 1 and comments about the French nuclear programme in Chapter 4 bear this out. Essentially the only countries where nuclear power is being built in any quantity are countries where health and safety standards and general labour markets allow lower construction costs compared to Western conditions. When these countries move towards Western conditions even the current modest rate of construction will tail off.

WHAT THE CORPORATIONS DO NOT WANT TO TALK ABOUT

One thing that corporations do not talk about (but they should) is 100 per cent renewable energy systems. Such systems (discussed in Chapter 1) are possible if (1) enough wind and solar is connected to an electricity grid and (2) there are appropriate balancing or storage systems and good grid connections. Longer-duration battery systems are being developed to ensure daily supply reliability (alongside demand-shifting services).

Long-term storage of renewable energy, either in the form of green hydrogen or green synthetic fuels, will also be needed. This can be stored in underground caverns – much in the same way as natural gas is stored at present. There is a trade-off here between (1) having 'overbuilding' of renewable energy so that there is much more renewable energy capacity available to supply demand at peak times and (2) having more long-term storage of renewable energy. There is, however, a near total silence about the 100 per cent renewable energy option. Perhaps that is not surprising since it would mean the complete restructuring, if not abolition, of the present corporate electricity regime. Wishful failure to understand the failures of nuclear power and fossil fuel CCS will lead to a dangerous position of failing to plan adequately to balance renewable energy.

WHY WE NEED A LOT MORE STATE INTERVENTION

Neoliberal efforts to mandate privatisation and artificial liberalisation schemes in areas which are natural monopolies showed up as spectacular failures during the European energy crisis (as discussed in Chapters 2 and 4). During the energy crisis the markets delivered ultra-high profits for energy corporations. Renewable energy was far cheaper in cost terms (as measured by the fixed price contracts with which they were developed) but their energy was often sold on for the high market price. As energy transition progresses this problem will continue in liberalised energy markets, with natural gas or substitutes (such as green hydrogen) setting a much higher umbrella market price on wholesale power markets. Energy generated by renewable energy will also be paid this high market price even though it is funded originally on a much cheaper basis because of fixed price contracts issued by governments.

A solution to these problems is more government intervention to set price caps on suppliers. This is most effectively managed through public ownership of the retail suppliers. This is in combination with contracts being issued by governments to renewable energy generators which do not reward the generators for high wholesale market prices. These contracts are called CfDs. Meaningful competition in the electricity sector can be maintained through competition in electricity generation between privately and publicly owned companies. State ownership can be diverse with ownership at municipal, regional and national levels.

There are four main reasons to have more state intervention to implement the energy transition. First is the enhancement of competition in the energy generation sector; second is to improve the ability of the energy system to balance fluctuating renewable energy; third is the protection of consumer interests, by ensuring lower costs of energy generation; and fourth is to develop local heat networks powered by large-scale heat pumps. This is discussed further in Chapter 5.

ENHANCEMENT OF COMPETITION THROUGH PUBLIC OWNERSHIP

The only sector of the energy industry that has possibilities for real competition is generation. Certainly, renewable energy began and thrived through independent generators starting up and then competing with centralised generation through fossil fuels and nuclear power. Increasingly, renewable energy is out-competing fossil fuels and (most obviously) new nuclear power in terms of cost. The degree to which renewable energy has been able to advance has been limited by the monopoly power of existing corporations and their influence on the incentives and regulatory system. In the USA, the energy corporations have often been able to limit the issuing of PPAs to renewable generators to protect their own established fossil fuel and nuclear power assets.

In the USA, the municipal electricity companies have until the recent past been frozen out of the possibilities for investment in renewable energy, because the latter have not been able to access the incentives. That is because the incentives have been reserved for the corporations who have the required tax base to claim the ITCs. However, now with the changes brought through in the IRA, which allow public authorities to access the incentives through cash, these publicly owned institutions can make direct investments in renewable energy. This increases the competition for renewable energy since we will now have more renewable energy developers in the generation market.

Governments should establish well-funded, publicly owned companies to invest in new renewable energy projects. Again, this increases competition in the renewable energy sector. More companies in the renewable energy sector produces greater competition. State-owned companies in Scandinavia are already delivering increasing quantities of well-organised, low-priced renewable energy projects.

HOW GOVERNMENT CONTRACTS FOR RENEWABLES WILL STOP PROFITEERING

The two things that are most needed for rapid renewable energy expansion are (1) the offer of lots of well-priced, long-term PPAs and (2) urgent action to build grid and distribution connections so that the projects can be developed. The provision of PPAs, organised by governments, can also have the effect of making sure that they are cost-effective for the consumer. Leaving things up to the corporations to organise renewable energy will end in more profiteering, with renewables replacing oil and gas as the source. What has been especially noticeable in European energy markets in the price crisis since late 2021 is that many companies have made what the economists euphemistically call 'economic rent' out of the situation. This includes the supply of fossil fuels, but it also involves supply of non-fossil fuels, whose costs should not vary with that of natural gas prices.

Oil and gas corporations have made large profits because of the high prices for fossil fuels in the 2021–3 period. In a 'marginal pricing' system, consumers can be ripped off by the energy corporations, as they sell otherwise cheap renewable energy at the same high prices as energy generated from gas-fired power plants. Marginal pricing systems are only acceptable when the consumers are protected by the existence of long-term PPAs between government agencies and the renewable energy generators.

We should not leave it to the big energy corporations to exploit renewable energy production for their own outrageous profit levels in the same way that they have exploited energy consumers with high fossil fuel prices. As discussed in Chapter 5, there has been a warning that, as in the case of a recent auction of offshore wind contracts in Germany, the corporations may control renewables production to extract as much profit as they can, rather than produce energy that is as cheap as possible for the consumer.

Government agencies should issue long-term PPAs to renewable energy projects through competitive tendering systems called CfDs. This system is widely used in the UK and ensures that the renewable energy generators are paid the same for each unit of energy produced, regardless of the level of wholesale electricity prices. The contract (CfD) price reflects what is needed, so that the developers can establish the projects and make a reasonable return on the investments in them. Corporations need to compete on equal terms with independent companies and publicly owned companies for government-backed generation contracts. Community-based projects could be awarded fixed price contracts using standardised prices. Currently, community schemes find it difficult to trade on large-scale power markets.

PUBLIC OWNERSHIP OF RETAIL ENERGY SUPPLY

Liberalised electricity supply markets, where a number of suppliers compete to serve domestic and medium-sized consumers, does not work very well, as discussed in the case of the UK in Chapter 2. There are too many consumers for suppliers to cost-effectively market themselves. The result is a poorly performing sham. Electricity supply (as opposed to generation) is a natural monopoly, except perhaps for the largest customers who have more resources to negotiate.

If retail electricity supply companies are state owned (either nationalised or municipally owned) then the cost savings from cheap renewable energy projects, organised through such contracting, will go straight through to consumers. Hence a publicly owned electricity supply can help consumers by limiting the costs that they have to pay. This is as opposed to excess profits being earned by energy corporations, as happens at times (such as 2021–3) when energy prices are very high because of natural gas price spikes.

It will not cost much to nationalise or municipalise the often almost bankrupt retail energy suppliers. Energy generation itself should be a competitive market, with publicly and privately owned companies competing to develop renewable energy projects funded as a result of the CfD and other contracts issued by governments.

Public ownership of the retail electricity supply will benefit the task of balancing fluctuating renewable energy. Competing domestic suppliers will not have a sufficiently big interest in spending a lot of money to incentivise consumers to balance the grid, for example to install home batteries. The suppliers have their own company interests in mind rather than the electricity system. However, if domestic supply is publicly owned then the system can be balanced in the interests of all consumers.

Transmission and distribution are also natural monopolies. However, they may cost more (than supply) to put into public ownership if they are not already in public hands.

PLANNING FOR ENERGY EFFICIENT HEAT NETWORKS

An especially effective way of using heat pumps is to deliver the heat from large-scale heat pumps through heat networks. Not all buildings can be connected to heat pumps, but a large proportion of urban buildings can be connected. Municipal authorities, which in any case oversee housing policies, are ideally suited local agents who can carry out this work.

In some countries (such as Germany, discussed in Chapter 5), the machinery for doing this already exists in the form of locally owned electricity companies. In Denmark (as mentioned in Chapter 4), the large bulk of heating is already delivered by district heating owned either by municipalities or cooperatives. Municipal authorities need to be set ambitious targets to deliver heat networks linked to large-scale heat pumps. In the UK, for example, the powers of local government have been hollowed out so much that they cannot raise enough capital to do the work, a situation that needs to be reversed.

Municipal authorities are also well suited to work with electricity companies to improve the energy efficiency of existing buildings through insulation programmes. A partnership between electricity companies and municipal authorities can also deliver a roll-out of heat pumps to individual buildings where heat networks (supplied by large-scale heat pumps) are not feasible. Funding of this programme can be done either through public spending or a levy on everybody's energy bills.

ENERGY DEMOCRACY

The battle to save the planet is a collective enterprise which needs energy democracy. It is not something that can merely be hived off to a few specialist, technocratic agencies. It needs investments to be redirected from fossil fuels into renewable energy and a complete change in the technology that is used to deliver the services that energy provides. Moreover, there is a need to build a green energy economy that avoids the profiteering that has been emblematic of the contemporary fossil fuel industries. This turn to energy democracy is discussed in Chapter 5.

Energy democracy includes the involvement of ordinary people and diverse interest groups in the decision-making process. As discussed in Chapter 5, there are different ways this can be achieved – through deliberative processes involving citizens making informed choices about policies, and also by establishing debates steered by independent experts. Certainly, this is an advance on the domination of the policy advice process by the corporations. Corporations will have access to the consultative processes like everyone else, and ministers are left to make the final policy choices. At present, industrial energy corporations may often be the only interests to have a major influence on policy outcomes. In this situation the corporations will steer policy towards their own interests rather than the interests of the consumer or the energy transition in general.

This energy democracy requires a diversity of types of ownership, replete with community participation at all levels and types. This includes community energy cooperatives, community solar and wind farms and shared community ownership of conventional wind and solar farms. This also includes an explosion of decentralised energy in buildings themselves through fitting solar panels and batteries. It involves DSM through effective system balancing organised by a publicly owned retail electricity supply. Corporate control of local energy distribution and supply can be municipalised – arguably, existing municipal ownership of energy systems represents a form of local energy democracy.

It also requires systems to involve ordinary people in making informed judgements as part of the policy process, and includes energy justice (as mentioned in Chapter 5) involving help to poorer communities in the energy transition from fossil fuels.

More generally, the costs of energy transformation need to be equitable, which means that the costs of the transition need to be spread around. This means that if people face high costs in altering their systems to be carbon neutral, then they need appropriate financial support do so. Above all, we should ignore those shameful 'whataboutery' pronouncements from politicians and corporations about how we ought to wait for China to take action. The developed nations, with their wealth and technological expertise, need to take the lead. After all, if the West had not initially promoted wind and solar power, developing nations would not now be deploying the technologies.

In short, developed nations must lead the way towards a 100 per cent renewable and energy efficiency revolution involving a lot more state intervention and energy democracy. An absence of energy democracy is likely to result in a failure to stop climate change and expensive dinosaur technologies for which energy consumers pay dearly. Energy democracy is likely to lead to lower energy prices, citizen involvement and a more rapid and equitable energy transition. Technology-focused social movements are desperately needed to organise a push towards these objectives. Only then shall we have less profiteering and more democracy, and moreover be in a good position to meet the climate goals set by the Paris Agreement.

Notes

PREFACE

1. IRENA (2023) 'World Energy Transitions Outlook 2023', International Renewable Energy Agency, page 56, www.irena.org/Publications/2023/Jun/World-Energy-Transitions-Outlook-2023.

CHAPTER 1

1. Quote reproduced originally in Toke, D. (2011) *Ecological Modernisation and Renewable Energy*, London: Palgrave, p. 7.
2. Data from Digest of UK Energy Statistics, www.gov.uk/government/collections/digest-of-uk-energy-statistics-dukes.
3. Data from US Energy Information Administration, www.eia.gov/.
4. Data drawn from Energy Institute Statistical Review of World Energy, 2023, www.energyinst.org/statistical-review/home?gclid=CjoKCQjwi7GnBhDXARIsAFLvH4mn2K13sMNYy3SSfsPumIJawF14xYiz-8m9JTSaamGE3KHwe1onVW8aAtLKEALw_wcB.
5. Data drawn from Energy Institute Statistical Review of World Energy, 2023, www.energyinst.org/statistical-review/home?gclid=CjoKCQjwi7GnBhDXARIsAFLvH4mn2K13sMNYy3SSfsPumIJawF14xYiz-8m9JTSaamGE3KHwe1onVW8aAtLKEALw_wcB.
6. Padovani, L. (2022) 'New Record as Wind and Solar Power Account for Nearly 60 per cent of Denmark's Annual Electricity Consumption', *Copenhagen Post*, 30 December, https://cphpost.dk/2022-12-30/news/new-record-as-wind-and-solar-power-account-for-close-to-60-percent-of-denmarks-annual-energy-consumption/.
7. For a detailed discussion of the Danish wind power movement in its early days, see Toke, D. (2011) 'Ecological Modernisation, Social Movements and Renewable Energy', *Environmental Politics*, 20, pp. 60–77.
8. Toke, D. (2007) *Making the UK Renewables Programme Fitter*, London: World Future Council, www.researchgate.net/profile/David-Toke/publication/242531945_Making_the_UK_Renewables_Programme_FITTER/links/00b49536bb5119b2ba000000/Making-the-UK-Renewables-Programme-FITTER.pdf.

9. Atani, M. (2019) 'Urgent Action Needed to Address Biomass in Africa, Says UNEP Study', UNEP, 19 November, www.unep.org/news-and-stories/press-release/urgent-action-needed-address-biomass-africa-says-unep-study.

10. Ivanova, I. (2023) '4 Oil Companies Had Total Sales of a Trillion Dollars Last Year', *CBS News*, 2 February, www.cbsnews.com/news/exxon-chevron-shell-conocophillips-record-profits-earnings-oil-companies-most-profitable-year/.

11. Carrington, D. (2022) 'Revealed: Oil Sector's "Staggering" $3bn-a-day Profits for last 50 Years', *The Guardian*, 21 July, www.theguardian.com/environment/2022/jul/21/revealed-oil-sectors-staggering-profits-last-50-years. See also Verbruggen, A. (2022) 'The Geopolitics of Trillion US$ Oil and Gas Rents', *International Journal of Sustainable Energy Planning and Management*, 36, https://doi.org/10.54337/ijsepm.7395.

12. Bell, S., Su, J., Kasper, M., Green S. and Keveke, C. (2023) 'Powerless in the United States: How Utilities Drive Shutoffs and Energy Injustice', Center for Biological Diversity, Energy and Policy Institute, BailoutWatch, www.biologicaldiversity.org/programs/energy-justice/pdfs/Powerless-in-the-US_Report.pdf.

13. ExxonMobil (2023) 'ExxonMobil Global Outlook: Our View to 2050', ExxonMobil, https://corporate.exxonmobil.com/-/media/global/files/global-outlook/2023/2023-global-outlook-executive-summary.pdf.

14. Muller, P. (1986) 'Transportation and Urban Form: Stages in the Spatial Evolution of the American Metropolis', in S. Hanson (ed.) *The Geography of Urban Transportation*, New York: Guilford Press, p. 60.

15. See for example YouTube video at www.youtube.com/watch?app=desktop&v=pUgVCR_658M.

16. Lovins, A. (1976) 'Energy Strategy: The Road Not Taken?', *Foreign Affairs*, 55(1).

17. Toke, D. (2023) 'Heat Pumps Will Save More Energy Than Insulation Says Top Energy Efficiency Expert', 100percentrenewableuk.org, 30 April, https://100percentrenewableuk.org/heat-pumps-are-more-important-than-insulation-for-saving-energy-says-top-energy-efficiency-expert.

18. Rosenow, J. and Hamels, S. (2023) 'Where to Meet on Heat? A Conceptual Framework for Optimising Demand Reduction and Decarbonised Heat Supply', *Energy Research and Social Science*, 104, p. 7.

19. Hausfather, Z. (2019), 'Factcheck: How Electric Vehicles Help to Tackle Climate Change', *Carbon Brief*, 13 May, www.carbonbrief.org/factcheck-how-electric-vehicles-help-to-tackle-climate-change/.

20. Harmon, J. (2023) 'New Design for Lithium-Air Battery Could Offer Much Longer Driving Range Compared with the Lithium-Ion Battery', Argonne Laboratory, 22 February, www.anl.gov/article/new-design-for-lithiumair-battery-could-offer-much-longer-driving-range-compared-with-the-lithiumion.

21. Patridge, J. (2021) 'Electric Vehicles Will Be Cheaper to Produce than Fossil Cars by 2027', *The Guardian*, 9 May, https://amp.theguardian.com/business/2021/may/09/electric-cars-will-be-cheaper-to-produce-than-fossil-fuel-vehicles-by-2027.

22. Nguyen, M. (2023) 'Innovation in EVs Seen Denting Copper Demand Growth Potential', Reuters, 10 June, www.reuters.com/business/autos-transportation/innovation-evs-seen-denting-copper-demand-growth-potential-2023-07-07/.

23. IEA Global EV Outlook (2023) 'Demand for Electric Cars Is Booming, with Sales Expected to Leap 35% This Year after a Record-Breaking 2022', IEA, www.iea.org/news/demand-for-electric-cars-is-booming-with-sales-expected-to-leap-35-this-year-after-a-record-breaking-2022.

24. Smil, V. (2005) *Energy at the Crossroads*, Cambridge, MA: MIT Press, pp. 348–59.

25. Lazard Consulting (2021) 'Levelised Cost of Energy, Levelized Cost of Storage, Levelized Cost of Hydrogen', Lazard, www.lazard.com/research-insights/levelized-cost-of-energy-levelized-cost-of-storage-and-levelized-cost-of-hydrogen-2021/.

26. Schneider, M. (2022) 'World Nuclear Industry Status Report 2022', Michael Schneider Consulting, www.worldnuclearreport.org/IMG/pdf/wnisr2022-v3-hr.pdf.

27. Green, J. (2023) 'Is Nuclear Power in a Global Death Spiral?' RENEW Economy, 3 March, https://reneweconomy.com.au/is-nuclear-power-in-a-global-death-spiral/#google_vignette.

28. Hanley, S. (2023), 'The Energy Technology Revolution Will Drive Renewable Energy Prices Even Lower', Cleantechnica, 1 September, https://cleantechnica.com/2023/09/01/the-energy-technology-revolution-will-drive-renewable-energy-prices-even-lower/.

29. World Nuclear Association (2023) 'Is the Cooling of Power Plants a Constraint on the Future of Nuclear Power?', https://world-nuclear.org/our-association/publications/technical-positions/cooling-of-power-plants.aspx.

30. Hunt, T. (2020) 'Has the International Energy Agency Finally Improved at Forecasting Solar Growth?', *PVMagazine*, 12 July, https://pv-magazine-usa.com/2020/07/12/has-the-international-energy-agency-finally-improved-at-forecasting-solar-growth/.

31. Fuller, G. (2014) 'Big Six Exploit Consumers Who Don't Switch', *The Times*, 16 December, www.thetimes.co.uk/article/big-six-exploit-customers-who-dont-switch-5n5rpnd9zlm.

32. Vanberg, V. (2004) *The Freiburg School: Walter Eucken and Ordoliberalism*, Freiburg: Institut für Allgemeine Wirtschaftsforschung, p. 10, cited in Toke D. and Lauber V. (2007) 'Anglo Saxon and German Approaches to Neo-liberalism and Environmental Policy: The Case of Financing Renewable Energy', *Geoforum*, 38(4), pp. 677–87.

33. Toke D. and Lauber V. (2007) 'Anglo Saxon and German Approaches to Neo-liberalism and Environmental Policy: The Case of Financing Renewable Energy', *Geoforum*, 38(4), pp. 677–87, p. 679.

34. Jackman, J. (2023) 'Which Countries Are Winning the European Heat Pump Race?', The EcoExperts, www.theecoexperts.co.uk/heat-pumps/top-countries.

35. Here I borrow a phrase used originally by Michel Foucault. See Garland, D. (2014) 'What Is a "History of the Present"? On Foucault's Genealogies and Their Critical Preconditions', *Punishment & Society*, 16(4), pp. 365–84.

36. See discussion about the influence of the radioactivity debate in Lutts, R. (1985) 'Chemical Fallout: Rachel Carson's Silent Spring, Radioactive Fallout, and the Environmental Movement', *Environmental Review*, 9, pp. 211–25.

37. Nye, D. (1998), *Consuming Power*, Cambridge, MA: MIT Press, p. 212.

38. Abdulla, A. et al. (2021) 'Explaining Successful and Failed Investments in U.S. Carbon Capture and Storage Using Empirical and Expert Assessments', *Environmental Research Letters*, 16, https://iopscience.iop.org/article/10.1088/1748-9326/abd19e/meta.

39. Howarth, R. and Jacobson, M. (2021) 'How Green Is Blue Hydrogen?', *Energy Science and Engineering*, 9(10), pp. 1676–87.

40. Brown, K. (2023) 'Why Do Construction Projects Cost So Much?' Maxim Recruitment, www.maximrecruitment.com/news/post/why-do-construction-projects-cost-so-much/.

41. Thomas, S. (2023) 'Small Modular Reactors: The Last Chance Saloon for the Nuclear Industry?', *Responsible Science*, 5, www.sgr.org.uk/sites/default/files/2023-05/SGR_RS5_2023_Thomas.pdf.

42. Australian Electricity Market Operator (2022) 'AEMO Unleashes ""Roadmap" towards 100% Renewables', 1 December, https://aemo.com.au/newsroom/media-release/engineering-framework-roadmap-to-100-per-cent-renewables.

43. Steitz, C., Alkousaa, R. and Sheahan, M. (2022) 'Germany Aims to Get 100% of Energy from Renewable Sources by 2035', Reuters, 27 February, www.reuters.com/business/energy/germany-step-up-plans-cut-dependence-russia-gas-2022-02-27/.

44. Jacobson, M. (2023) *No Miracles Needed: How Today's Technology Can Save Our Climate and Clean Our Air*, Cambridge: Cambridge University Press.

45. Jacobson, M. (2021) *100% Clean, Renewable Energy and Storage for Everything*, Cambridge: Cambridge University Press, p. 176.

46. Blakers, A. (2023) 'Despairing about Climate Change? These 4 Charts on the Unstoppable Growth of Solar May Change Your Mind', *The Conversation*, 11 May, chart 'Annual wind + solar generation per person', https://theconversation.com/despairing-about-climate-change-these-4-charts-on-the-unstoppable-growth-of-solar-may-change-your-mind-204901?utm_source=twitter&utm_medium=bylinetwitterbutton#.

47. Vorath, S. (2023) 'South Australia Hits Stunning New High in Race to Renewables-Only Grid', Renew Economy, 11 January, https://reneweconomy.com.au/south-australia-hits-stunning-new-high-in-race-to-renewables-only-grid/.

48. Parkinson, G. (2022) 'South Australia Remarkable 100 Per Cent Renewable Run Extends to Over Ten Days', Renew Economy, 22 December, https://reneweconomy.com.au/south-australias-remarkable-100-per-cent-renewables-run-extends-to-over-10-days/.

49. Jacobson, M. (2021) *100% Clean, Renewable Energy and Storage for Everything*, Cambridge: Cambridge University Press, p. 176.

50. Jacobson, M. (2023) *No Miracles Needed: How Today's Technology Can Save Our Climate and Clean Our Air*, Cambridge: Cambridge University Press, p. 253, assuming 8.7 TW of world power needed in 2050 as stated in Jacobson, M. (2021) *100% Clean, Renewable Energy and Storage for Everything*, Cambridge: Cambridge University Press.

51. BVG-Wind Europe (2017) *Unleashing Europe's Offshore Wind Potential*, Brussels: Wind Europe, p. 29, assuming UK electricity consumption of 320 TWh per year, https://windeurope.org/wp-content/uploads/files/about-wind/reports/Unleashing-Europes-offshore-wind-potential.pdf.

52. Schneider, M. (2022) 'World Nuclear Industry Status Report 2022', pp. 16 and 284, www.worldnuclearreport.org/IMG/pdf/wnisr2022-v3-lr.pdf.
53. Fragaki, A., Markvart,T. and Laskos , G. (2019) 'All UK Electricity Supplied by Wind and Photovoltaics: The 30–30 Rule', *Energy*, 169, pp. 228–237, p. 231.
54. Nature Portfolio (2022) 'Hydrogen Gas Turbine Offers Promise of Clean Electricity', www.nature.com/articles/d42473-022-00211-0.
55. Eavor (2022), 'Enabling Energy Autonomy ... Everywhere', www.eavor.com/.
56. See discussion in Diesing, P., Bogdanov, D., Satymov, R., Child, M. and Breyer, C. (2023) '100 Per Cent Renewable Energy for the United Kingdom', 100percentrenewableuk.org, https://100percentrenewableuk.org/wp-content/uploads/100-RE-23-Dec-.pdf.
57. See for example relative cost of power plant discussed in a report by Diesing, P., Bogdanov, D., Satymov, R., Child, M. and Breyer, C. (2023) '100 Per Cent Renewable Energy for the United Kingdom', 100percentrenewableuk.org, https://100percentrenewableuk.org/wp-content/uploads/100-RE-23-Dec-.pdf.
58. Wu, C., Zhang, X. and Sterling, M. (2021) 'Global Electricity Interconnection with 100% Renewable Energy Generation', Institute of Electrical and Electronic Engineers, https://ieeexplore.ieee.org/document/9511456.

CHAPTER 2

1. Kinnock, N. (1992) House of Commons debate, 19 October, Hansard Column 210, https://publications.parliament.uk/pa/cm199293/cmhansrd/1992-10-19/Debate-1.html.
2. Friends of the Earth Press Release (2022) 'Govt's Climate Strategy Deemed "Unlawful" in Historic Ruling', Friends of the Earth, 22 July, https://friendsoftheearth.uk/climate/govts-climate-strategy-deemed-unlawful-historic-ruling.
3. Jolly, J. and Sweeney, M. (2022) 'Big Oil's Quarterly Profits Hit £50bn as UK Braces for Even Higher Energy Bills', *The Guardian*, 2 August, www.theguardian.com/business/2022/aug/02/big-oil-profits-energy-bills-windfall-tax.
4. Jarvis, C. (2022), 'Energy Company Profits: the Staggering Sums Pocketed by Firms While Our Energy Bills Soar', *Left Foot Forward*, 15 August, https://leftfootforward.org/2022/08/energy-company-profits-the-staggering-sums-pocketed-by-firms-while-

our-bills-soar/; Thomas, N. (2022) 'SSE Pledges to Reinvest Windfall Profits in UK Energy Assets', *Financial Times*, 27 September, www. ft.com/content/b329b16f-401c-48d7-a5f8-da36d68f9de7.

5. Millward, R. (1991) 'The Market Behaviour of Local Utilities in Pre-World War I Britain: The Case of Gas', *The Economic History Review*, 44(1), pp. 102–27; Coase, R. (1950) 'The Nationalisation of Electricity Supply in Great Britain', *Land Economics*, 26(1), pp. 1–16.

6. Millward, R. (1991) 'The Market Behaviour of Local Utilities in Pre-World War I Britain: The Case of Gas', *The Economic History Review*, 44(1), pp. 102–127; Coase, R. (1950) 'The Nationalisation of Electricity Supply in Great Britain', *Land Economics*, 26(1), pp. 1–16.

7. Burchell, G. (1996) 'Liberal Government and Techniques of the Self', in A. Barry, T. Osborne and N. Rose (eds), *Foucault and Political Reason*, Abingdon: Routledge, pp. 22–3.

8. Edwards, T. (2022) Cornwall Insight, comment on Twitter, 6 October, https://twitter.com/NotionalGrid/status/1577964112121892864.

9. Department of Trade and Industry (1995) 'The Prospects for Nuclear Power in the UK', Cm2860, para 3.38, p. 16.

10. Cipan, V. (2022) 'European Gas Storage Reserves: By Country Update Daily', 13 October, https://viborc.com/europe-gas-storage-reserves-capacities-by-country.

11. OFGEM (2022) 'Large Legacy Suppliers: Domestic Fuel Bill Breakdown over Time', retail market indicators, www.ofgem.gov.uk/energy-data-and-research/data-portal/retail-market-indicators.

12. Digest of United Kingdom Energy Statistics (2023) Energy Trends, https://assets.publishing.service.gov.uk/government/uploads/system/uploads/attachment_data/file/1147239/Energy_Trends_March_2023.pdf.

13. Institute of Energy Statistical Review of World Energy, 2023.

14. Lawson, A. (2023) 'Jeremy Hunt Accused of "£20bn" Gamble on Nuclear Energy and Carbon Capture', *The Guardian*, 15 March, www. theguardian.com/uk-news/2023/mar/15/jeremy-hunt-accused-of-taking-a-20bn-gamble-on-nuclear-energy-and-carbon-capture.

15. Environment at a Glance Indicators (2023) 'Climate Change', OECD iLibrary, www.oecd-ilibrary.org/sites/5584ad47-en/index. html?itemId=/content/component/5584ad47-en.

16. Blakers, A. (2023) 'Despairing about Climate Change? These 4 Charts on the Unstoppable Growth of Solar May Change Your Mind', *The Conversation*, 11 May, chart 'Annual wind + solar generation per person', https://theconversation.com/despairing-about-climate-change-these-4-charts-on-the-unstoppable-growth-of-solar-may-

change-your-mind-204901?utm_source=twitter&utm_medium=
bylinetwitterbutton.

17. Nicol. S., Roys, M., Ormandy, D. and Ezratty, V. (2016) 'Poor housing
in the European Union', BRE, p. 10, https://files.bregroup.com/
bre-co-uk-file-library-copy/filelibrary/Briefing papers/92993_BRE_
Poor-Housing_in_-Europe.pdf; Tado (2020) 'UK Homes Losing
Heat up to Three Times Faster than European Neighbours', www.
tado.com/gb-en/press/uk-homes-losing-heat-up-to-three-times-
faster-than-european-neighbours.

18. Mullane, J. (2023) 'Why Is the UK Lagging Behind in Its Heat
Pump Uptake while the Rest of the World Steams Ahead?' *Home-
building and Renovating*, 26 April, www.homebuilding.co.uk/news/
why-is-britain-lagging-behind-in-its-heat-pump-uptake.

19. Mortimer-Lee, P. and Mao, X. (2022) 'Deindustrialisation in the UK',
National Institute UK Economic Outlook, Spring 2022, www.niesr.
ac.uk/wp-content/uploads/2022/05/Deindustrialisation-in-the-UK.
pdf.

20. ENTRANZE (2023) Average Floor Area per capita, https://entranze.
enerdata.net/.

21. US Energy Information Agency (2022) 'Today in Energy', 22
February, www.eia.gov/todayinenergy/detail.php?id=51358.

22. National Audit Office (2022) 'The Energy Supplier Market', Session
2022-23, 22 June, HC 68, www.nao.BULBorg.uk/wp-content/
uploads/2022/03/The-energy-supplier-market.pdf.

23. O'Mahony, M. and Vecchi, M. (2001), 'The Electricity Supply
Industry: A Study of an Industry in Transition', *National Institute
Economic Review*, 177, pp. 85–99.

24. Piketty, T. (2013) *Capital in the Twenty-First Century*, Cambridge,
MA: Harvard University Press.

25. Lawrence, M. (2022) 'Power to the People: The Case for a
Publicly Owned Generation Company', Commonwealth, www.
common-wealth.co.uk/reports/power-to-the-people-the-case-for-a-
publicly-owned-generation-company.

26. See Diesing, P., Bogdanov, D., Satymov, R., Child, M. and Breyer, C.
(2023) '100 Per Cent Renewable Energy for the UK', 100percentre-
newableuk.org, https://100percentrenewableuk.org/wp-content/
uploads/100-RE-23-Dec-.pdf

27. Plimmer, G. (2022) 'UK Bills Inflated by Failure to Implement EU
Power Cables Deal', *Financial Times*, 29 September, www.ft.com/
content/19616e10-4991-4986-a555-6e14605340f1.

28. TUC (2022) 'A Fairer Energy System for Families and the Climate', July 2022, www.tuc.org.uk/research-analysis/reports/fairer-energy-system-families-and-climate.
29. UNITE Union (2023) 'UNITE Investigates: Renationalising Energy, Costs and Savings – Full Report', www.unitetheunion.org/what-we-do/unite-investigates/unplugging-energy-profiteers-the-case-for-public-ownership/unite-investigates-renationalising-energy-costs-and-savings-full-report#.
30. Wilson, J. (2019) 'The NHS as a Monopsony', *BMJ Opinion*, September, https://blogs.bmj.com/bmj/2019/09/20/jacob-wilson-the-nhs-as-a-monopsony.
31. Labour Party (2022) 'Keir Starmer Calls for New National Champion in Clean Energy, Great British Energy with a Mission to Cut Bills, Create Jobs and Deliver Energy Independence', 27 September, https://labour.org.uk/press/keir-starmer-calls-for-new-national-champion-in-clean-energy-great-british-energy-with-a-mission-to-cut-bills-create-jobs-and-deliver-energy-independence/.
32. Welsh Government Press Release (2022) 'Wales Announced Publicly Owned Renewable Energy Developer', 25 October, https://gov.wales/wales-announces-publicly-owned-renewable-energy-developer.
33. Commonweal (2020) 'Economy, Energy and Fair Work Committee Publicly Owned Energy Company Inquiry Submission from Common Weal', https://archive2021.parliament.scot/S5_Economy-JobsFairWork/Inquiries/EEFW-S5-18-POEC-9-CommonWeal.pdf.
34. Willis, R., Mitchell, C., Hoggett, R., Britton, J., Poulter, H., Pownall, T. and Lowes, R. (2019) 'Getting Energy Governance Right: Lessons from IGov', University of Exeter/EPSRC, http://projects.exeter.ac.uk/igov/wp-content/uploads/2019/08/IGov-Getting-energy-governance-right-Sept2019.pdf.

CHAPTER 3

1. Limb, L. (2023) 'Major Milestone for EU Energy: Wind and Solar Produced More Electricity than Gas in 2022', euronews.green, 9 January, www.euronews.com/green/2023/01/31/major-milestone-for-eu-energy-wind-and-solar-produced-more-electricity-than-gas-in-2022.
2. EIA (Energy Information Service) (2023) 'Increasing Renewables Likely to Reduce Coal and Natural Gas Generation over Next Two Years', www.eia.gov/todayinenergy/detail.php.

3. Enerdata (2023) 'Germany's Power Consumption Falls in 2022, Generation from Renewable Energy Rises', 5 January, www. enerdata.net/publications/daily-energy-news/germanys-power-consumption-falls-2022-generation-renewables-rises.html.

4. Calculated using data derived from Windfair (2023) 'Wind and Solar Photovoltaic Electricity Generation Break Records in Spain in 2022', 9 January, https://w3.windfair.net/wind-energy/pr/42836-red-electrica-spain-grid-renewable-energy-generation-technology-solar-wind-electricity.

5. Padovani, L. (2022) 'New Record As Wind and Solar Power Account for Close to 60 Percent of Denmark's Annual Electricity Consumption', *Copenhagen Post*, 30 December, https://cphpost.dk/2022-12-30/news/new-record-as-wind-and-solar-power-account-for-close-to-60-percent-of-denmarks-annual-energy-consumption/.

6. Spacik, V. (2023) 'Solar, Wind Power almost All Household Electricity Use in China', *Balkan Green Energy News*, 20 February, https://balkangreenenergynews.com/solar-wind-power-almost-all-households-in-china-in-2022/.

7. OECD (2013) 'Climate', www.oecd-ilibrary.org/sites/5584ad47-en/index.html?itemId=/content/component/5584ad47-en.

8. Bourgeois, A., Lafrogne-Joussier, R., Lequien, M. and Ralle, P. (2022) 'One Third of the European Union's Carbon Footprint Is Due to Its Imports', *INSEE Analysis*, 74, 20 July, www.insee.fr/en/statistiques/6478761.

9. Energy Information Administration (2023) 'Use of Explained', www.eia.gov/energyexplained/use-of-energy/#.

10. Global Cooling Prize (2021) 'Breakthrough, Climate-Friendly ACs: Winners of the Global Cooling Prize Announced', 29 April' https://globalcoolingprize.org/grand-winners-press-release/.

11. Elliason, M. (2020) 'Seattle Must Require Public Buildings Meet Passivhaus Standards to Lead on Climate', The Urbanist, July 22, www.theurbanist.org/2020/07/22/seattle-must-require-passivhaus/.

12. Steinbauer, J. (2021) 'Gas Industry Fuels a Legislative Campaign to Keep Its Fires Burning', Sierra, 8 July, www.sierraclub.org/sierra/gas-industry-fuels-legislative-campaign-keep-its-fires-burning.

13. Steinbauer, J. (2021) 'Gas Industry Fuels a Legislative Campaign to Keep Its Fires Burning, Sierra, 8 July, www.sierraclub.org/sierra/gas-industry-fuels-legislative-campaign-keep-its-fires-burning.

14. NYSERDA (2023) 'US Heat Pump Sales Surpass Gas Furnaces', New York State, www.nyserda.ny.gov/Featured-Stories/US-Heat-Pump-Sales.

15. Stein, Z. (2023) 'Investor Owned Utilities (IUOs)', Carbon Collective, 24 March, www.carboncollective.co/sustainable-investing/investor-owned-utilities-ious.
16. Energy Information Administration (2023) 'Where Solar Is Found and Used', www.eia.gov/energyexplained/solar/where-solar-is-found.php.
17. For a comparison of wind resources in different US states see Office of Energy Efficiency and Renewable Energy (2023), 'Wind Exchange', https://windexchange.energy.gov/maps-data/325.
18. Quote taken from Toke, D. (2011) *Ecological Modernisation and Renewable Energy*, Basingstoke: Palgrave Macmillan, p. 122.
19. Toke, D. (2011) *Ecological Modernisation and Renewable Energy*, Basingstoke: Palgrave Macmillan, p. 123.
20. Energy Information Administration (2023) 'Wyoming: State Profile and Energy Estimates', www.eia.gov/state/analysis.php?sid=WY.
21. Wind Exchange (2023) Office of Energy Efficiency and Renewable Energy, US Government, https://windexchange.energy.gov/projects/tax-credits.
22. Gipe, P. (2016) *Wind Energy for the Rest of Us*, Bakersfield, CA: WindWorks, p. 466.
23. Knuth, S. (2022) 'Renewable Energy: US Tax Credits for Wind and Solar Mostly Benefit Big Banks', *The Conversation*, 25 January, https://theconversation.com/renewable-energy-us-tax-credits-for-wind-and-solar-mostly-benefit-big-banks-173965.
24. Knuth, S. (2022) 'Rentiers of the Low-Carbon Economy? Renewable Energy's Extractive Fiscal Geographies', *Environment and Planning A*, open access, https://journals.sagepub.com/doi/10.1177/0308518X211062601.
25. Knuth, S. (2022) 'Renewable Energy: US Tax Credits for Wind and Solar Mostly Benefit Big Banks', *The Conversation*, 25 January, https://theconversation.com/renewable-energy-us-tax-credits-for-wind-and-solar-mostly-benefit-big-banks-173965.
26. Toke, D. (2023) 'Government to Pump Billions of Subsidies to Fossil Fuel Industry to Support Failed Technology', 100percentrenewableuk.org, https://100percentrenewableuk.org/government-to-pump-billions-of-subsides-to-fossil-fuel-industry-to-support-failed-technology.
27. Weston, D. (2015) 'Cape Wind Threatened by Utilities over PPAs', *Wind Power Monthly*, 7 January, www.windpowermonthly.com/article/1328234/cape-wind-threatened-utilities-ppas.
28. See for example Miller, J. (2021) 'The Texas Public Utility Commission's Revolving between Industry and Regulator', *San*

Antonio Current, 5 March, www.Sacurrent.Com/News/the-texas-public-utility-commissions-revolving-door-between-industry-and-regulator-25708910; Avalos, G. (2015) 'PUC Critics Cite Concerns over "Revolving Doors"', *Mercury News*, 21 March, www.mercurynews.com/2015/03/21/puc-critics-cite-concerns-over-revolving-door/.

29. Hlinka, M. (2021) 'Utility Commissioners: Who They Are and How They Impact Regulation', S&P Market Intelligence, 11 May, www.spglobal.com/marketintelligence/en/news-insights/blog/us-utility-commissioners-who-they-are-and-how-they-impact-regulation.

30. McGowan, E. (2018) 'Virginia Ruling Points to Bumps in the Road for State's Renewable Mandate', *Energy Network News*, 7 June, https://energynews.us/2018/06/07/virginia-ruling-points-to-bumps-in-the-road-for-states-renewable-mandate/.

31. Justia US Law (2014), *SZ Enters, LLC v. Iowa Utils. Bd.*, https://law.justia.com/cases/iowa/supreme-court/2014/130642.html.

32. Eller, D. (2018) 'Power Struggle: Iowa Town Takes on Utility Giant for Right to Go Greener', *Des Moines Register*, 19 April, https://eu.desmoinesregister.com/story/money/business/2018/04/19/alliant-energy-power-decorah-electric-utility-vote-referendum-midamerican-iowa-energy-green-solar/430145002/.

33. Unsigned (2022), 'Mandate versus Movement: State Public Service Commissions and Their Evolving Power over Our Energy Source', *Harvard Law Review*, 35, 11 April, https://harvardlawreview.org/2022/04/mandate-versus-movement/.

34. Stokes, L. (2020) *Short Circuiting Policy*, New York: Oxford University Press, p. 90.

35. Stokes, L. (2020) *Short Circuiting Policy*, New York: Oxford University Press, pp. 214–223.

36. Stokes, L. (2020) *Short Circuiting Policy*, New York: Oxford University Press, pp. 133–140.

37. Busby, J. et al. (2021) 'Cascading Risks: Understanding the 2021 Winter Blackout in Texas', *Energy Research & Social Science*, 77, 102106.

38. McClaughlin, T. (2022) 'Creaky US Power Grid Threatens Progress in Renewables, EVs', Reuters, 12 May, www.reuters.com/investigates/special-report/usa-renewables-electric-grid/.

39. Kennedy, R. (2023) 'California Utilities Commission Rejects Solar Microgrid Proposal', *PV Magazine*, 13 April, https://pv-magazine-usa.com/2023/04/13/california-utilities-commission-rejects-solar-microgrid-proposal/.

40. Toke, D. (2018) *Low Carbon Politics*, Abingdon: Routledge, p. 106.

41. Frye, R. (2008) 'The Current "Nuclear Renaissance" in the United States, Its Underlying Reasons, and Its Potential Pitfalls', *Energy Law Journal*, 29(2), pp. 279–9, p. 283.

42. Larson, A. (2021) 'Former CEO Will Land in Prison for Two Years as a Result of VC Summer Nuclear Project', *Power*, 15 October, www.powermag.com/former-scana-ceo-will-land-in-prison-as-result-of-v-c-summer-nuclear-project/.

43. AP Press (2022), 'Georgia Nuclear Plants Cost Now to Top $30 Billion', *GPB News*, 9 May, www.gpb.org/news/2022/05/09/georgia-nuclear-plants-cost-now-forecast-top-30-billion; Gardner, T. (2019) 'U.S. Finalizes $3.7 Billion Loan for Vogtle Nuclear Power Plant', Reuters, 22 March, www.reuters.com/article/us-usa-nuclearpower-vogtle-idUSKCN1R31X9.

44. Landers, M. (2021), 'What Ratepayers Should Know about the Vogtle Expansion', *The Current*, 4 January, https://thecurrent ga.org/2022/01/04/what-ratepayers-should-know-about-the-vogtle-expansion/.

45. Lyman, E. (2022) 'Coalition for Responsible Energy Development in New Brunswick (CRED-NB)', CBC interview, 22 April, https://crednb.ca/2022/04/12/ed-lyman-cbc-interview-on-smrs/.

46. Welch, D. (2023) 'How Much Do Solar Panels Cost in 2022?', Green.org, 21 February, https://green.org/2022/10/26/how-much-do-solar-panels-cost-in-2022/.

47. Woody, T. (2012) 'Cutting the Price of Solar in Half by Cutting Red Tape', *Forbes Magazine*, 12 July, www.forbes.com/sites/toddwoody/2012/07/05/cut-the-price-of-solar-in-half-by-cutting-red-tape/?sh=2ff4d3e8495e.

48. Yoder, K. (2023) 'Utilities Use Consumer Dollars to Pay for Lobbying: Here's How Lawmakers Can Stop it', *Grist*, 26 January, https://grist.org/regulation/utilities-lobbying-corruption-climate-change-report/.

49. Zientara, B. (2021) 'The State of Net Metering in the United States in 2021', *Solar Reviews*, 21 January, www.solarreviews.com/blog/the-state-of-net-metering-usa-2021.

50. Database for Incentives for Renewable Energy (DSIRE) (2021), 'Important Information Regarding Third Party PPAs for Solar Installations', August 2021, https://ncsolarcen-prod.s3.amazonaws.com/wp-content/uploads/2021/12/DSIRE_3rd-Party-PPA_Aug_2021-2.pdf.

51. Ayres, A. and Wheeles, D. (2022) 'How California Can Expand Solar Development and Support San Joaquin Valley Farmers', *CAL*

Matters, 27 October, https://calmatters.org/environment/2022/10/california-water-drought-farm-solar-development/.

52. Editorial (2023) 'California Needs a Lot More Solar: Why Not Put Panels along Highways and Parking Lots?', *LA Times*, 16 May, www.latimes.com/opinion/story/2023-05-16/la-ed-solar-highways-parking-lots.

53. Georghiu, I. (2022) 'Gov. Desantis Vetoes Rooftop Solar Bill, Citing Desire to Not Add to "Financial Crunch" Facing Floridians', Utility Dive, 22 April, www.utilitydive.com/news/Florida-desantis-vetoes-rooftop-solar-bill-behind-the-meter-cost-shifting/622820.

54. Caplan, A. (2022) 'Florida Power and Light Tied to Dark Money Used to Help Sen. Keith Perry Win 2018 Race', *Gainsville Sun*, 14 August, https://eu.gainesville.com/story/news/2022/08/14/florida-power-light-dark-money-2018-state-senate-race/10277503002/.

55. Dinan, T. (2017), 'Federal Support for Developing, Producing and Using Fuels and Energy Technologies', Congressional Budget Office, Testimony to House of Representatives Sub-Committee, 29 March, www.cbo.gov/system/files/115th-congress-2017-2018/reports/52521-energytestimony.pdf.

56. Dinan, T. (2017) 'Federal Support for Developing, Producing and Using Fuels and Energy Technologies', Congressional Budget Office, Testimony to House of Representatives Sub-Committee, 29 March, p. 12, www.cbo.gov/system/files/115th-congress-2017-2018/reports/52521-energytestimony.pdf.

57. Energy Information Administration (2021) 'Most Planned US Natural Gas Fired Plants Are in Appalachia and in Florida and Texas', 22 November, www.eia.gov/todayinenergy/detail.php?id=50436.

58. Satterfield, M. (2022), 'Freeport LNG Fire Illustrates the Inflationary Impact of LNG Exports', Industrial Energy Consumers of America, 14 June, www.ieca-us.com/wp-content/uploads/06.14.22_Freeport-Press-Release.pdf.

59. BP Statistical Review of World Energy (2023), p. 11.

60. Environment at a Glance Indicators (2023) 'Climate Change', OECD iLibrary, www.oecd-ilibrary.org/sites/5584ad47-en/index.html?itemId=/content/component/5584ad47-en.

61. Leahy, S. (2019) 'This is the World's Most Destructive Oil Operation – and It's Growing', *National Geographic*, 11 April, www.nationalgeographic.com/environment/article/alberta-canadas-tar-sands-is-growing-but-indigenous-people-fight-back?loggedin=true&rnd=1686074700466.

62. Wilson Centre (2012) 'So Canada left Kyoto: Why? And What's next?', 12 March, www.wilsoncenter.org/event/so-canada-left-kyoto-why-and-whats-next; Lo, J. (2021) 'Oil and SUVs: Why Canada's Emissions Have Risen Since Trudeau Took Office', *Climate Home News*, www.climatechangenews.com/2021/09/16/oil-suvs-canadas-emissions-risen-since-trudeau-took-office/.

63. Chopson, P. (2021) 'Energy Codes at the Provincial Level', 15 December, https://cove.tools/blog/canadian-energy-codes-province-2021.

64. Fairlie, I. (2008) 'The Hazards of Tritium: Revisited', *Medicine Conflict and Survival*, 24, pp. 306–19, https://pubmed.ncbi.nlm.nih.gov/19065871/.

65. Crawley, M. (2017) 'How Privatised Power Haunts Ontario Politics', *CBC News*, 9 December, www.cbc.ca/news/canada/toronto/ontario-hydro-bills-privatization-1.4439500.

66. Email correspondence with Jack Gibbins, Ontario Clean Air Alliance, 6 June 2023.

67. Provincial and Territorial Energy Profiles, Quebec (2021), Canada Energy Regulator, www.cer-rec.gc.ca/en/data-analysis/energy-markets/provincial-territorial-energy-profiles/provincial-territorial-energy-profiles-quebec.html.

68. Urban, R. (2021), 'Electricity Prices in Canada', Energy Hub, www.energyhub.org/electricity-prices/.

CHAPTER 4

1. Unless otherwise stated energy data are drawn from the BP Statistical Review of World Energy, 2022.

2. Kates, G. and Luo, L. (2014) 'Russian gas: How Much Is That?', *Radio Free Europe*, 1 July, www.rferl.org/a/russian-gas-how-much-gazprom/25442003.html.

3. Graf, A., Gagnebin, M. and Buc, M., 'Breaking Free from Fossil Gas', *Agora-Energiewende*, https://static.agora-energiewende.de/fileadmin/Projekte/2021/2021_07_EU_GEXIT/A-EW_292_Breaking_free_WEB.pdf.

4. Energy Information Administration (2022) 'Three Countries Provided almost 70% of Liquified Natural Gas Received in Europe in 2021', www.eia.gov/todayinenergy/detail.php.

5. EU Commission (2023), Renewable Energy Directive, Energy Directorate, https://energy.ec.europa.eu/topics/renewable-energy/

renewable-energy-directive-targets-and-rules/renewable-energy-directive_en.

6. Toke, D. (2008) 'The EU Renewables Directive: What Is the Fuss About Trading?', *Energy Policy*, 36, pp. 3001–8. www.sciencedirect.com/science/article/abs/pii/S0301421508001869?via%3Dihub.

7. Toke, D. (2011) *Ecological Modernisation and Renewable Energy*, Basingstoke: Palgrave Macmillan.

8. Rankin, J. (2019) 'Eastern European Countries Block EU Moves Towards 2050 Net Zero Target', *The Guardian*, 20 June, www.theguardian.com/environment/2019/jun/20/eu-leaders-to-spar-over-zero-carbon-pledge-for-2050.

9. Eichhammer, W. and Chlechowitz, M. (2021) 'Does the EU Emission Trading Scheme ETS Promote Energy Efficiency?', ODYSSEE-MURE project, www.odyssee-mure.eu/publications/policy-brief/ets-promote-energy-efficiency.pdf.

10. Embacher, J. and Thomas, S. (2023) 'Electricity Market Reform: Take "the Market" out', *Social Europe*, 23 April, www.socialeurope.eu/electricity-market-reform-take-the-market-out.

11. EPSU (2022) 'Going Public: A Decarbonised, Affordable and Democratic Energy System for Europe – The Failure of Energy Liberalisation', European Federation of Public Service Unions, www.epsu.org/sites/default/files/article/files/Going%20Public_EPSU-PSIRU%20Report%202019%20-%20EN.pdf.

12. Johansen, K. and Werner, S. (2022) 'Something Is Sustainable in the State of Denmark: A Review of the Danish Heating Network', *Renewable and Sustainable Energy Reviews*, 158, pp. 112–17, www.sciencedirect.com/science/article/pii/S1364032122000466.

13. Leidreiter, A. (2013) 'Denmark Puts the Brake on Heating Costs with New Legislation', *Renewable Energy World*, 15 February, www.renewableenergyworld.com/blog/denmark-puts-the-brakes-on-heating-costs-with-new-legislation/#gref.

14. Blakers, A. (2023) 'Despairing About Climate Change? These 4 Charts on the Unstoppable Growth of Solar May Change Your Mind', *The Conversation*, 11 May, https://theconversation.com/despairing-about-climate-change-these-4-charts-on-the-unstoppable-growth-of-solar-may-change-your-mind-204901?utm_source=twitter&utm_medium=bylinetwitterbutton.

15. Unfortunately I have been unable to do charts showing changes in Danish energy consumption between 2011 and 2021 because the database that I am using, namely BP's Statistical Review of World

Energy, did not give relevant statistics for 2022 in the case of Denmark.

16. Vyhnak, J. (2020) 'Students Survival Guide: The Essentials about Cycling in Denmark', 8 February, www.studentsurvivalguide.dk/posts/all-the-essentials-about-cycling-in-denmark.

17. Toke, D. (2018) *Low Carbon Politics*, Abingdon: Routledge, pp. 69–71.

18. Toke, D. (2005) 'Are Green Electricity Certificates the Way Forward for Renewable Energy? An Evaluation of the UK's Renewables Obligation in the Context of International Comparisons', *Environmental Planning C*, 23(3), pp. 361–75.

19. Energdata (2023) 'Germany's Gas Consumption and Imports Declined in 2022', 3 January, www.enerdata.net/publications/daily-energy-news/germanys-gas-consumption-and-imports-declined-2022.html.

20. Keating, D. (2021) 'Why the German Greens Want to kill Nordstream 2', *Energy Monitor*, 14 May, www.energymonitor.ai/networks-grids/why-the-german-greens-want-to-kill-nord-stream-2/.

21. Lepesant, G. (2023) 'Higher Renewable Energy Targets in Germany: How Will Industry Benefit?', French Institute of International Relations (IFRI), www.ifri.org/en/publications/briefings-de-lifri/higher-renewable-energy-targets-germany-how-will-industry-benefit.

22. Unsigned (2022) '10 GW of New Wind Farms a Year: German Parliament Adopts New Onshore Wind Law', *WindEurope*, 11 July, https://windeurope.org/newsroom/news/10-gw-of-new-wind-farms-a-year-german-parliament-adopts-new-onshore-wind-law/.

23. Jasper, J. (1990) *Nuclear Politics*, Princeton, NJ: Princeton University Press, p. 94.

24. Jasper, J. (1990) *Nuclear Politics*, Princeton, NJ: Princeton University Press, pp. 94–95.

25. Kitschelt, H. (1986) 'Political Opportunity Structures and Political Protest: Anti-nuclear Movements in Four Democracies', *British Journal of Political Science*, 16(1), pp. 57–85.

26. Liberatore, A. (1999) *The Management of Uncertainty: Lessons from Chernobyl*, Abingdon: Routledge, p. 165.

27. Liberatore, A. (1999) *The Management of Uncertainty: Lessons from Chernobyl*, Abingdon: Routledge, pp. 173–96.

28. Toke, D. (2021) *Nuclear Power in Stagnation*, Abingdon: Routledge, p. 100.

29. Mouterde, P. (2023) 'France to Pay up to €500 for Falling Short of Renewable Energy Targets', *Le Monde*, 14 June, www.lemonde.fr/

en/environment/article/2022/11/25/renewable-energy-france-will-have-to-pay-several-hundred-million-euros-for-falling-short-of-its-objectives_6005566_114.html.

30. Unsigned (2023) 'EDF Poses Record Loss in France due to Reactor Outages', *World Nuclear News*, 17 February, www.world-nuclear-news. org/Articles/EDF-posts-record-loss-in-France-due-to-reactor-out; Enerdata (2023) 'EDF's Power Generation in France Reached a Record Low in 2022', 17 January www.enerdata.net/publications/ daily-energy-news/edfs-power-generation-france-reached-record-low-2022.html.

31. Crellin, F. and Eckart, V. (2022) 'French Nuclear Woes Stoke Europe's Power Prices', *Financial Times*, 24 August, www.reuters. com/business/energy/french-nuclear-woes-stoke-europes-power-prices-2022-08-24/.

32. Guillet, J. (2022) 'EDF's Woes Are a Bigger Long Term Problem for EU Energy than the War in Ukraine', Jerome A. Paris, 24 July, https:// jeromeaparis.substack.com/p/edfs-woes-are-a-bigger-long-term.

33. Toke, D. (2021), *Nuclear Power in Stagnation*, Abingdon: Routledge, p. 102, quotes from interviews conducted on 9 September and 19 March.

34. See Guillet, J. (2022) 'EDF's Woes Are a Bigger Long Term Problem for EU Energy than the War in Ukraine', Jerome A. Paris, 24 July, https:// jeromeaparis.substack.com/p/edfs-woes-are-a-bigger-long-term

CHAPTER 5

1. Judson, E., Fitch-Roy, O. and Soutar, I. (2022) 'Energy Democracy: A Digital Future?', *Energy Research and Social Science*, 91, 102732.

2. Vote Solar Priorities (2023), Vote Solar, https://votesolar.org/ policy-priorities/.

3. Mountain West (2023), Vote Solar, https://votesolar.org/mountain-west/.

4. Warmuth, J. (2023) 'Vote Solar Celebrates Passage of Historic Clean Energy Legislation', 3 February, https://votesolar.org/ vote-solar-celebrates-passage-of-historic-clean-energy-legislation/.

5. Delurey, J. (2021) 'Climate and Equitable Jobs Act Passes Illinois Legislature', Vote Solar, 13 September, https://votesolar.org/ climate-and-equitable-jobs-act-passes-illinois-legislature/.

6. Possible Impact Prospectus (2023) 'Our Proudest Achievements', p. 3, https://static1.squarespace.com/static/5d30896202a18c0001b4

9180/t/64087080aa7b90219618c18e/1678274768918/Possible+
impact+prospectus+-+2023.pdf.

7. We Are Possible (2023) 'Our Projects', www.wearepossible.org/
current-projects.

8. Mardell, S. and Richardson, J. (2022) 'Supporting Coal Workers
and Communities in the Energy Transition', Rocky Mountain
Institute, 15 September, https://rmi.org/supporting-coal-workers-
and-communities-in-the-energy-transition/.

9. Denzel, T., Diehl, D. and Henninger, J (2023) 'Heizen Wie Heidelberg', Tagesschau, 16 June, www.tagesschau.de/inland/innenpolitik/
heizung-kommunen-waermeplanung-100.html.

10. Schlandt, J. (2015) 'Small but Powerful: Germany's Municipal Utilities', *Clean Energy Wire*, 18 February, www.cleanenergywire.org/
factsheets/small-powerful-germanys-municipal-utilities.

11. Toke, D. (2011) 'Ecological Modernisation, Social Movements and
Renewable Energy', *Environmental Politics*, 20, pp. 60–77, p. 73.

12. Provost, C. and Kennard, M. (2014) 'Hamburg at Forefront of
Global Effort to Reverse Privatisation of Local Services', *The
Guardian*, 12 November, www.theguardian.com/cities/2014/nov/12/
hamburg-global-reverse-privatisation-city-services.

13. Sellier, T. (2023), 'APPA Statement on Notice of Proposed Rulemaking
for Elective Payments under the Inflation Reduction Act', American
Public Power Association, 14 June, www.publicpower.org/
publication/appa-statement-notice-proposed-rulemaking-elective-
payments-under-inflation-reduction-act-ira.

14. Nassar, H.-M. (2023) 'Vancouver Isn't Banning Gas Stoves, Fire-
places in New Builds', *CityNews*, 1 June, https://vancouver.citynews.
ca/2023/06/01/vancouver-natural-gas-ban-changes/.

15. Alpack-Ripka, L. and Penn, I. (2020) 'How Coal-Loving Australia
became the Leader in Rooftop Solar', *New York Times*, 1 October,
www.nytimes.com/2020/09/29/business/energy-environment/
australia-rooftop-solar-coal.html.

16. Penn, I. (2022) 'Hit Hard by High Energy Costs Hawaii Looks to
the Sun', *New York Times*, 30 May, www.nytimes.com/2022/05/30/
business/hawaii-solar-energy.html.

17. Hawaiin Public Utilities Commission (2023) 'Performance Based
Regulation', https://puc.hawaii.gov/energy/pbr/.

18. Belgian Ministry of Energy (2022) 'Public Consultation on the
Offshore Wind Tender for the Princess Elisabeth Zone', 19 January,
p. 23, https://economie.fgov.be/sites/default/files/Files/Energy/

public-consultation-on-the-offshore-wind-tender-for-the-princess-elisabethzone.pdf.

19. SeaCoop (2023) 'Citizen Offshore Power', https://seacoop.be/en/citizen-offshore-power/.

20. Solar Energy Industries Association (SEIA) 2023, 'Community Solar', www.seia.org/initiatives/community-solar; Wood Mackenzie (2023) 'US Community Solar Growth Slowed 16% in 2022: National Market Expected to Double by 2027', www.woodmac.com/press-releases/us-community-solar-growth-slowed-16-in-2022-national-market-expected-to-double-by-2027/.

21. Shared Solar (2023) Community Energy Projects, www.sharedrenewables.org/community-energy-projects/.

22. Rapin, C. (2023) 'Local Residents in Puerto Rico Build the Island's First Community Owned Solar Microgrid', *Renewable Energy World*, 12 June, www.renewableenergyworld.com/solar/microgrids-solar/local-residents-in-puerto-rico-built-the-islands-first-community-owned-solar-microgrid/.

23. Dutta, R. (2023), 'Local Solar Offers the Lowest Cost Pathway to 100% Clean Energy', *PVMagazine*, 21 June, https://pv-magazine-usa.com/2023/06/21/local-solar-offers-the-lowest-cost-pathway-to-100-clean-energy/.

24. Kennedy, R. (2023) 'California Utilities Commission Rejects Solar Microgrid Proposal', *PVMagazine*, 13 April, https://pv-magazine-usa.com/2023/04/13/california-utilities-commission-rejects-solar-microgrid-proposal/.

25. Willis, R., Curato, N. and Smith, G. (2022) 'Deliberative Democracy and the Climate Crisis', *Wiley Interdisciplinary Reviews*, 11 January, p. 3, https://wires.onlinelibrary.wiley.com/doi/epdf/10.1002/wcc.759.

26. Summary of description made in Wachtel, T. (2019) 'How a Citizen's Legislature Made Texas #1 in Renewable Energy', Climate Emergency UK, 14 July, www.climateemergency.uk/blog/how-a-citizens-legislature-made-texas-1-in-renewable-energy/.

27. Wachtel, T. (2019) 'How a Citizen's Legislature Made Texas #1 in Renewable Energy', Climate Emergency UK, 14 July, www.climateemergency.uk/blog/how-a-citizens-legislature-made-texas-1-in-renewable-energy/.

28. UK Parliament (2023), 'Climate Assembly UK', www.parliament.uk/get-involved/committees/climate-assembly-uk/.

29. Cherry, C. E., Capstick, S., Demski, C., Mellier, C., Stone, L. and Verfuerth, C. (2021) *Citizens' Climate Assemblies: Understanding Public*

Deliberation for Climate Policy, Cardiff: Centre for Climate Change and Social Transformations, p. 32.

30. Cherry, C. E., Capstick, S., Demski, C., Mellier, C., Stone, L. and Verfuerth, C. (2021) *Citizens' Climate Assemblies: Understanding Public Deliberation for Climate Policy*, Cardiff: Centre for Climate Change and Social Transformations, p. 32.

31. Coleman M., Delaney, L., Torney, D. and Bereton, P. (2019) 'Ireland's World-Leading Climate Assembly, What Worked? What Didn't?', *Climate Home News*, 27 July, www.climatechangenews.com/2019/06/27/irelands-world-leading-citizens-climate-assembly-worked-didnt/.

32. Toke, D. and Nielsen, H. O. (2015) 'Political Consultation and Political Styles: Renewable Energy Consultations in the UK and Denmark', *British Politics*, 10(4), pp. 454–44, pp. 460–1.

33. Durakovic, A. (2023) 'Breaking: Germany Rakes in EUR12.6 Billion through "Dynamic" Bidding Offshore Wind Auction', OffshoreWIND.biz, 12 July, www.offshorewind.biz/2023/07/12/breaking-germany-rakes-in-eur-12-6-billion-through-dynamic-bidding-offshore-wind-auction/.

34. Guillet, J. (2023) 'Why the German Tender for Offshore Wind is Bad News', Jerome A. Paris, 14 July, https://jeromeaparis.substack.com/p/why-the-german-tender-for-offshore?utm_source=post-email-title&publication_id=583570&post_id=134822657&isFreemail=true&utm_medium=email

35. Van Veelen, B. and van der Horst, D. (2018) 'What Is Energy Democracy? Connecting Social Science Energy Research and Political Theory', *Energy Research and Social Science*, 46, pp. 19–28.

Index

Thanks to our Patreon subscriber:

Ciaran Kane

Who has shown generosity and
comradeship in support of our publishing.

Check out the other perks you get by subscribing
to our Patreon – visit patreon.com/plutopress.

Subscriptions start from £3 a month.

The Pluto Press Newsletter

Hello friend of Pluto!

Want to stay on top of the best radical books
we publish?

Then sign up to be the first to hear about our
new books, as well as special events,
podcasts and videos.

You'll also get 50% off your first order with us
when you sign up.

Come and join us!

Go to bit.ly/PlutoNewsletter